AF841246

A PENNY TO NELSON'S PILLAR

Portrait of Kyran Quaney, Electrical Engineer and Inventor, referred to in the pages of this book as Kerr.

A PENNY TO NELSON'S PILLAR

by Joseph Quaney

Volturna Press

© Joseph Quaney 1971

First Published November 1971

Printed and published by Volturna Press
Portlaw, Co. Waterford, Eire.

Also: Dunmore, Pitlochry, Scotland;
Sancho Dávila 32, Madrid 2, Spain;
(New Cosmopolitan Book Service)
2901 Webster Street, San Francisco,
California 94123

Illustrations printed by Design-Print Ltd.,
Waterford, Eire.

SBN 85606 017 8

Introduction

THE ROMANCE of the Roads has never been written. In early times, the onward march of civilisation, depending as it did on roads, slowed up and virtually stopped, with the decline and fall of the Roman Empire. Centuries later, when King George the Third of England was caught up in the whirligig of the Industrial revolution, he concentrated on building roads throughout Great Britain and Ireland — to open up new trade routes. During his reign hundreds of "Turnpike Acts" were passed and the cost of maintaining the roads was met by tolls, collected at toll-gates. At their best, the roads in those days were rocky, uneven and often marshy and impassable.

It was not until the year 1815 that John Loudon MacAdam, a Scottish engineer, invented a system of road building which became generally adopted. The new highways were called macadam roads after the name of the inventor. These modern roads literally paved the way for the era of the horse and the first public system of horse-drawn trams started in Paris in 1819. Most of the early horse-drawn trams were run on the surface of the road without rails.

Horse trams on rails made their first appearance in Britain when an American named George F. Train, opened a passenger tramcar service in Birkenhead and three lines in London in 1860. These were successful and after this experiment similar services started all over Europe.

There was little by way of organised public transport in rural Ireland, until 1815 when an Italian, named Charles Bianconi started a two-wheel post-chaise service; to carry passengers and mails around Clonmel, in County Tipperary. This venture became popular and successful and the service was extended — with the exception of Dublin and Belfast. The original coaches were changed to four-wheel "Long Toms"; capable of carrying ten passengers. By 1845, Bianconi's post chaises ran on over 3,000 miles of Irish roads. When the railways took over delivery of the mails in 1850, Bianconi, faced with a big loss in revenue, disposed of his post-chaise enterprise and invested his capital in railway shares.

The new railway and canal services brought untold prosperity to Dublin around mid-century. Industries boomed and the population increased to such an extent that overcrowding became a major problem. Rival transport groups were unable to cope with the crowds until William Martin Murphy, a financial genius, united them and formed a horse-tram monopoly to meet Dublin's passenger transport requirements.

This eased the situation for years, until at last people began to realise that the horse, as a public transport unit, was unable to keep pace with the times. Demands for a modern rapid-transit service grew in strength.

At this critical stage, William Martin Murphy announced that after discussing the matter with electrical engineers, he could with their help and technical experience provide an electric tramcar service second to none in Europe.

The story of how this was accomplished is now related in the following pages.

Joseph Quaney

electrical engineers, he could with their help and technical experience provide an electric tramcar service second to none in Europe.

The story of how this was accomplished is now related in the following pages.

1

THE DUANE

storey farmhouse at Bawngorte

miles from Roscrea in the hea

Snugly set on a hill-side, its g

work and bulgy whitewashed w

against a dark background of t

crowded the tiny windows, da

little was known about the whole

light in the year 1870.

A dark-brown turf rick, flan

walled outhouses, nestled again

a five-acre field at the back, a

ed noisily through the stubble

led and squawked around the kit

On this clear autumn morning

sented as pretty a picture of pea

as one could wish to see, obvio

good ancestral line, in which eac

justified its existence by enrich

passing it on to the next.

To the north and west lay dre

but the aspect to the south was

1

THE QUANEYS lived in a long one-storey farmhouse at Bawngorra, a townland twelve miles from Roscrea in the heart of rural Ireland. Snugly set on a hill-side, its golden thatch, red paint-work and bulgy whitewashed walls, stood out prettily against a dark background of trees. Scarlet geraniums crowded the tiny windows, darkening the interior, but little was known about the wholesome effects of sun-light in the year 1870.

A dark brown turf rick, flanked by a huddle of mud-walled outhouses, nestled against the gable end. In a five-acre field at the back, a flock of geese search-ed noisily through the stubble and barnyard fowls cack-led and squawked around the kitchen door.

On this clear autumn morning, the homestead pre-sented as pretty a picture of peace and contentment as one could wish to see, obviously the holding of a good ancestral line, in which each generation had justified its existence by enriching the heritage before passing it on to the next.

To the north and west lay dreary stretches of bog, but the aspect to the south was fair and fertile, the land gently undulating as far as the eye could see to the blue-grey foothills of the Slieve Bloom mountains. Knockaroo church, with its square-cut tower and white flecked graveyard, stood out on a rise in the middle distance and its deep-toned bell, mingling with the rural sounds of peasantry and kine, seemed to invoke a blessing on the peaceful scene. Whether it tolled for a wedding, for the Angelus, or for the peace of a departed soul, the church bell had a powerful influence on the local community, gently reminding them of their Faith and their dependence on Providence for the fruits of the earth.

Gilbert Quaney was a bit of an oddity. He loved books and took little interest in his farm beyond gaining a living from it. People said he had missed his vocation and that he should have been a schoolmaster or a priest. On a Sunday morning, wearing surcoat and cravat and sporting a silver-knobbed stick, he would have passed for either. After Mass he would drop in on a neighbour or two for a friendly chat.

Gilbert was considered an authority on heraldry and folklore. While dinner was cooking he would entertain the family, tracing back their lineage for generations, and perhaps re-tell a story about the crazy escapades of some ancestor. In those easy-going times there was always a welcome, and a drop in the bottle for a congenial soul who could while away an hour or two in agreeable conversation.

Life followed a simple course in the remote farmhouse at Bawngorra, especially during the long winter evenings. When rush lights gleamed and threw fantastic shadows on the walls, and the turn fire glowed upon the kitchen hearth, the family would foregather to talk and listen to the wind howling like a banshee in the darkness outside. The children would snuggle close together, shielding their eyes and spluttering when a furious gust rattled the windows and sent smoke billowing down the chimney and on up until it reached the blackened rafters.

Margaret, in her forties, careworn from the endless drudgery which is the lot of a farmer's wife, would clear up after the evening meal or perhaps bake bread for the next day in a bastable by the fire.

Kerr, the eldest son, a wiry dark-haired lad of ten, and Pat, a plump and slobbery gosoon two years his junior, usually shared the battered armchair while Mollie, Julia and Maggie, ages ranging from four to seven, struggled for best place on the low settle. Little Maggie, as yet unable to hold her own, invariably wound up sprawling tearfully on the hard earthen floor.

Divileens the childer could be at times, and when Gilbert came in after feeding the stock and stood, back to the fire, wisps of steam rising from his clothes as he readied his stubby black pipe, Margaret would call on him to chide or punish those who had misbehaved during the day. Then a hush would fall on everyone as he gave out, or pretended to look for the ash plant; they feared his temper, which was fiery when aroused.

The little domestic storms soon blew over however, and by the time Gilbert got his pipe going and at last settled down in the armchair, respectfully vacated by the boys, the girleens would be giggling again and the gosoons annoying him with questions, as intelligent lads of their age usually do.

Uncle Michael, weary after his hard day's work, would smoke his pipe in the inglenook and perhaps interpose a word or two. He was very fond of the children and they all loved him. So the evenings passed pleasantly enough, the youngsters gradually forming their first ties of kinship and a natural love for home and fireside.

If it was the month of October, Margaret would produce her beads and all would kneel down to say the Rosary. She would begin the prayers herself and each of the men folk would say a decade in turn. Then, after the final chorus of discordant but none the less reverent 'Amens', the family would retire to bed.

One evening in November, Gilbert and Uncle Michael returned from the fair in Borris-in-Ossory bringing with them two loaves of baker's bread for the "childer". This was a change from the everyday home-made bread and was considered a great luxury in those days. They also brought a tortoiseshell comb as a peace offering for Margaret, who was always upset and grieved to see them with the signs of drink.

The brothers were full, in every sense of the word, and in the best of form after a good market in town and when the children settled down after supper, Gilbert, in an expansive mood, announced that the time had come to tell them something about their family history.

He carried their minds back to a period of civil and religious strife in Europe, known as the Huguenot wars, which were raging when Catherine de Medici was Queen Regent of France. At that time, a direct ancestor of theirs, François Quane, held a position of high honour as military attaché to the Queen. One day, while waiting at the Tuileries for the Court to assemble, François was introduced to James Maurice Fitzgerald, an Irishman who had journeyed to France to enlist sympathy and help for his country's cause.

Fitzgerald's plan was to organise a fighting division which would sail to Ireland and give military aid to an insurrection which was already planned by powerful Irish chieftans.

Catherine, beset by her enemies the Huguenots and flaunted by her renegade son King Henry the Third, who favoured an alliance with England, realised that a rebellion in Ireland would cause a political upheaval in London which would weaken the English position in France, gave Royal assent to the plot, and appointed

François Quane to organise and lead the French expeditionary force which ultimately landed at Dingle, in Kerry, in the year 1579.

News of the landing spread and soon all Munster was seething with rebellion. The O'Donoughues of Kerry, the O'Briens of Limerick and the McCarthys of Cork rallied to the call and the army marched eastward like a rising tide, gathering force and reinforcements on the way.

Waterford, the English stronghold, was the first objective and 'Death to the invader' the slogan. The spirit was truly fanatical but the revolt was badly planned. It was not long before this became evident. Food supplies and equipment were inadequate and long marches in bad weather took a heavy toll of casualties before any fighting started. By the time the insurgents joined forces with Fitzgerald's cousin, the rebel Earl of Desmond, near Tallow in Waterford, the hastily recruited legion was in a sorry plight.

Spies carried this information to Sir Walter Raleigh, who, without giving them time to recuperate or reorganise, marched out from Lismore castle at the head of a powerful well-disciplined army to engage them. He at once attacked the French division, the spearhead of the insurrection, and inflicted heavy losses.

James Maurice Fitzgerald was mortally wounded and his death had a demoralising effect on his followers. The Earl of Desmond's forces were massacred with a ferocity seldom equalled on the battlefield. Within four days, François was forced to withdraw, to save his division from annihilation.

They fought a rearguard action along the road they had travelled full of high hopes a week before, mercilessly harassed by the English dragoons, until their orderly retreat became a rout. A Council of War was held and it was decided that the remainder of the expedition would disband so that all would have an equal chance to escape and survive.

Accordingly they split up into groups and struck out across the country, taking refuge in monasteries, hamlets and villages far from the battlefield. When it became known that they had fought for Ireland, the Frenchmen were hospitably received and encouraged to settle down wherever they chose to stay.

Thrown on their own resources, they soon adapted themselves to their new surroundings and set about starting life anew. They founded industries such as silk weaving, glass blowing and lead mining all of which prospered. Masons, carpenters and other craftsmen resumed their trades, introducing new architectural styles, and the result of their labours may be seen to this day in the distinctive groups of quaint Huguenot houses which perpetuate their memory in places as far apart as Tullamore, Portarlington and Dublin.

Some, like François, intermarried with the peasanttry and lived out their lives in the peace and seclusion of a farm.

By the time Gilbert reached this part of his narrative, Uncle Michael was snoring in the inglenook; Mollie, Julia and little Maggie were fast asleep, and Pat was so drowsy that he couldn't keep his eyes open. Kerr, however, was fascinated by the story. He listened, as the French would say, "with all his ears" until his father stood up a bit unsteadily and solemnly imparted the family tradition.

The gist of it was: "Be a decent and God-fearing man, work hard to make your way in life and don't be too easily swayed by the opinions of others."

Young Kerr, who looked up to his father as the most knowledgeable person in the world, was so impressed by the way he spoke that this maxim became the guiding influence in his career.

In December, 1880, the bell of Knockaroo church tolled for the passing of Gilbert, head of the house at

Bawngorra. He was a young man by the standard of a community noted for longevity; barely in his fifties.

It was whispered that his death was hastened by over-addiction to liquor and at his wake stories were told of how Rory, the big mastiff, used to knock him down when he had a drop too much, and hold him there, until he sobered up enough to walk properly. No one thought the worse of him for that, however; few of the community had been held in greater esteem. After the funeral, the family gathered in the big farmhouse parlour for the will to be read out by the solicitor.

When the first overwhelming tumult of her grief had passed, poor saintly Margaret became resigned to her loss and faced up to the future with courage and fortitude that showed true nobility of character. As none of the children were old enough to run the farm and she felt unable to cope with it, Uncle Michael, her brother-in-law, agreed to take over until one of the boys came of age. It was decided that Pat should be placed in the care of Master Clancy, who conducted the hedge school in Borris, and that Kerr would be sent to a boarding school in Dublin, as soon as the parish priest could make arrangements. Margaret would look after the girls for the time being.

At the age of ten, Kerr showed a natural bent for mechanical work. When other lads were playing "Peelers and Robbers," the local equivalent for "Cowboys and Indians," Kerr would be down in Flaherty's forge, irresistibly drawn by the smell of burning hoofs and the ring of Mick's three-pound hammer on the anvil.

He helped the smith to shoe horses or repair machinery, worked the bellows and held the tongs, fascinated by the sizzling metal and the rocketing sparks, learning every detail of the job.

"That gosoon'll make a master craftsman some day, I'd like to have him for a 'prentice," said Mick to

Margaret. But that was not to be. About six weeks after his father's death, Kerr bade tearful goodbye to all at Bawngorra and Uncle Michael took him in the trap to Ballybrophy, to catch the train for Dublin.

In the Sacred Heart Home in Drumcondra, he learned the rudiments of his spiritual and temporal education. At the end of each term he returned home, gladdening them all with his steady development and scholarly attainments. He was over thirteen and the parish priest, impressed by his school reports, hoped and prayed that one day he would follow in his footsteps and become a priest.

With this in mind he arranged that Kerr would undergo a course of study at Blackrock College, on the outskirts of Dublin, where young men were coached for the professions and a select few were encouraged to study for the Church.

Life at this seminary was austere, almost monastic, the teaching being carried out by a militant order of priests, whose code of discipline and standard of education was said to be the highest in Europe.

Kerr made good progress from the start of his college career. He excelled at physics and electrodynamics, the new science which attracted most of the brainy and ambitious students. The teachers rated him a genius.

Five years passed quickly. Kerr had three staunch friends, Frank, Leo and Tom, all like himself in their final term. Examinations were over and the usual atmosphere of suppressed excitement was evident as the Christmas holidays drew near. Farewell parties were held in the senior rooms, surreptitiously of course, but the Dean usually closed a tolerant eye to these festivities unless the participants became too hilarious.

One evening the four companions met in Kerr's room for a little celebration and all went smoothly until Frank produced a bottle of spirits. By the time it was empty, they were in form for any kind of devilment. Frank, whose Dad owned a pub in the city, suggested that they go to a quiet place he knew of and make an evening of it. They did.

When the party was returning over the college railings at two in the morning, as rowdy as moonlighters, Tom caught his foot between the spikes and sprained his ankle. He hung head downwards, moaning with pain. Suddenly a light appeared in the Headmaster's window and the sash was raised.

It was only a matter of hours until the janitor had three of them lined up in the Dean's study. Tom would have been there too, but he was too lame to walk.

As old Carey explained their misdoings of the previous night, the Headmaster's frown deepened to an expression of profound disgust. These young men had taken part in an escapade which might well bring the good name of the College into disrepute.

There was no excuse for Kerr. As a prefect it was his duty to show good example to others. He had broken his trust. The Dean's glance singled him out.

"As a result of this, your transfer to Maynooth College, which I had arranged for January, will be postponed indefinitely," he said icily. "The others may go until I decide how to deal with them. That is all."

Kerr felt he could never hold up his head again. When he returned home humiliated, Uncle Michael broke the news to Margaret. It was a heavy cross for her to bear. Her son a "spoiled priest," an object of pity to all!

They feared for his health and implored him to forget about careers and settle down on the farm which one day would become his own. It had prospered beyond their dreams and there was a worthwhile share for every member of the family. But Kerr's outlook on life had changed. Higher education and mixing with the sons of gentry had inspired him with ideals which could never be realised on a farm.

He had studied the lives of Edison, Alexander Graham Bell, Elias Howe and other famous inventors who had risen from obscurity by their own efforts, and was fired with an ambition to do likewise. After a long discussion it was decided that he would be given his share of the farm in cash. That would enable him to pay his fare to America and leave enough in hand to keep him going until he got a start in the New World.

2

SOFTLY outlined by a crescent moon, the 'Campania' lay-to off Queenstown, waiting for passengers. Thousands of lights reflected in the wavelets as tenders moved alongside and made fast to the vast, towering hull. Emigrants, carrying all their possessions, clambered through doors in the ship's side.

Kerr found his berth in cabin number thirty-four on the middle deck. Shortly afterwards, bells jangled overhead, chains rattled in the hawse pipes and cries of "All ashore that's goin' ashore" echoed through the corridors. Suddenly a deep-toned blast overwhelmed the minor sounds and the liner, pulsating from stem to stern set a course for NewYork.

After the first day at sea, when the business of unpacking and settling down in their new quarters was over, passengers who were not laid low with sea sickness began to pair off.

"Bit o' comfort in this outfit," said Packey, a cabin mate, as he and Kerr took seats in the dining room on the second day out. "Different to the one I sailed to Baltimore in, about thirty years ago. The

Nordic, they called her, a paddle-wheeler, used paddles in fine weather an' sails when it got too rough. There wasn't any arrangements for cookin' or eatin' at all. Everyone was supposed to bring their own grub; or buy it off the steward. You cooked it in th' galley — if y' could fight your way into it."

"Sounds real rough" said a fellow on his left.

"Rough it was an' no mistake," continued Packey, warming to his story. "Passengers was herded in th' 'tween decks like cattle, ne'r a table to eat on; men, women an' kids squattin' around on their luggage an' sleepin' wherever they could get their heads down.

"Some of 'em took sick an' died withoug seein' a doctor or a priest — kep' a stock o' coffins aboard for them. Cap'n read a bit o' th' Bible an' they was shoved over th' side without any more fuss. Battened the hatches down to keep 'em quiet at night. An' a right bloody mix-up it was when th' weather got bad," he continued. "One half o' th' crew was sailors an' t'other half firemen, 'Paddy Westers' they useta call 'em. Th' firemen didn't know a thing about sails an' the sailors knew nothin' about stokin' fires; but all was supposed to muck in with th' runnin' o' th' ship.

"Never ferget th' time the wind kep' shiftin' an' the crew was switchin' from steam to sail an' sail to steam all day. They was wore out. There was ructions when the mate called all hands on deck to wear ship.

"While the bos'un an' his mates was roustin' 'em out, the ship got caught in a squall an' heeled over on her beam end. There was a crash below as if the world had come to an end, as the coal in th' 'tween decks slid over to the lee side. Th' starboard paddle was buried in th' sea an' th ' one on th' port side swished around in a cloud of steam as if it had gone mad.

"Th' screams an' roars o' the passengers heaped on top o' one another in th' darkness below was terrible — everyone thought th' ship was goin' down. Howsomever," continued the speaker, "after wallowin' in the trough o' the sea all night, th' vessel steadied a bit, an' th' crew got a chance to trim the coals an' put her on an even keel again. Six o' th' passengers died an' a crowd of 'em was badly hurt. Don't know your luck travellin' on a packet like this," he concluded, as the steward served the after-dinner coffee.

Packey had roughed it most of his life. In his early years he had fought for the Confederates in the Civil War. He had worked his way all over the States and despite his grey hair, was an active and likeable character. Kerr learned a lot from his new friend.

The weather remained fine and the liner made good progress. One afternoon there was a flutter of excitement when the rumour spread that a landfall was expected. Everyone was on the lookout, competing for the honour of being the first to shout "Land Ho! " It was too late, however when Sandy Hook Light was spotted in the darkness; all but a few hardy passengers had gone below.

As the pilot was not due until dawn, fires were banked down and the liner hove to, pending his arrival. In the cabins everyone was on the move. Bags and trunks were hauled out from their hiding places, packed and un-packed over again. The hunt for passports and tickets was on. Stewards flitted up and down the corridors, collecting their harvest of tips.

Snatches of "Sweet A —deline" and "Auld Lang Syne" came from the bar long after closing time, until at last the chief steward appeared and persuaded the roisterers to break it up and turn in; so they could rise and sing like skylarks in the morning.

At four a.m. next day, Sandy Hook light lay a mile or so off to port.

"Haul down the 'Pilot Jack'," shouted the chief officer as a cutter bobbed alongside and a figure in oilskins clambered up the Jacob's ladder, over the bulwark and on up the iron steps to the bridge.

"Howdy, Cap'n," he introduced himself breezily, shaking hands. "Any fever aboard?"

"No," returned the captain. "Care for a tot, Pilot?" he enquired in an undertone.

"Not till we dock, thanks, Cap'n. Hope we make it on this tide, I'm due a week-end off."

Seagulls squawked around the stern.

"Ready for movin', Mr. Mate?" shouted the pilot.

"All clear forrid, Sir," returned the chief officer.

"O.K., Cap'n, shove her ahead at half speed; keep that red buoy on your port bow, helmsman. That's it son: steady as she goes."

The air was clear and cold as the liner steamed slowly through the Narrows, past Fort St. George. As the course changed, the vast panorama of the upper Bay swung into view with the Manhatten sky-line looming in the distance. Miles of wharves and shipyards lined the shores of the Bay.

Lofty square-rigged ships and barques were moored to tiers of buoys, waiting for cargo at Bayonne. Flood traffic was at its peak; ships of all classes and sizes furrowed the waters of the bay.

Outward bound with grain for London, pianos for Melbourne, locomotives for Argentina, or perhaps inward with silk from China, sugar from Demerara, spices from the Indies, or wines from Spanish ports; whalers, tankers, fruit boats, trawlers, minions of commerce, plying through dangerous shoals and storm-wracked seas, dauntless, weatherbeaten craft, pitting their strength and the skill of their crews against the inexorable forces of time and tide.

The 'Campania' moved at half-speed.

"Watch that son-of-a-bitch ahead!" roared the captain, wrenching the handle of the engine room telegraph as a big tramp steamer swung into the fairway.

"Thinks there's no one in th' goddam say but himself!" growled the pilot, giving a blast on the whistle. "That fella oughta be certified."

Slowly the mammoth structure of the Brooklyn Bridge came into view as they rounded a bend.

"Man who designed that never saw it finished," remarked the pilot. "Got his foot crushed between a ferry an' one of the piers an' died. Roebling was his name."

The immigrants swarmed on deck, ready for going ashore: women with bundles and babies, men grouping their families around piles of baggage. As the liner drew abreast of the Battery excitement spread.

Irish, Polacks, Italians and Jews pushed towards the rails, eager to get a better view of the magic metropolis where dreams were supposed to come true. In 1886 few of the buildings in the Manhattan sky-line were over three hundred feet high, but to the poverty-stricken immigrants who saw them for the first time, they were as thrilling and romantic as enchanted castles in Spain.

As the vessel moved towards dockland the noise of industry increased. A dredger groaned at its Herculean task. Dockyard hammers added to the din. Ferries hooted as they weaved through the traffic. Small craft scurried out of the way, buffeting the grey-flecked current as it swept along the waterfront out to sea.

The liner's crew stood at mooring stations. Tugs appeared at bow and stern and deftly moved the vessel to the wharf. The routine task of making fast was quickly accomplished and gangways were run out to entrances in the dock wall. The pilot and captain descended to the chartroom.

"Think I'll have one now, Cap'n," said the pilot, as the steward appeared with glasses on a tray. The passengers could no longer be restrained. With an irresistible surge, two colourful streams of humanity flowed down the gangways into the subways, stamping the ground with land-happy feet as they swept on to the Customs Barriers. When the irritating inspection of passports and luggage was over, they pushed on to the exits and out into the street, to be caught up, whirled along and lost in the mighty current of traffic in New York.

3

KERR followed the crowd to the station door where he had arranged to meet Packey, but his friend was nowhere to be seen. Bewildered by vast buildings, whirling traffic and hustling crowds, he gathered up his traps and darted across to the corner of Fourteenth Street.

He remembered Packey's instructions. "A few blocks down from the station an' turn south, anyone'll tell ya th' way to Irishtown."

He trudged along the dusty, sun-baked sidewalk, perspiring. At the corner of Ninth Avenue he dropped his bags for a breather.

A noise overhead caused him to raise his eyes as an elevated train thundered to a stop on the platform above. Seconds later people streamed down the iron stairways into the street. A coffee-coloured man fell over his bags and swore.

Suddenly a whistle shrilled, the engine belched smoke and with a wheezy gasp the train rattled off into the distance. A waggon team panicked and bolted, scattering market stalls, ash cans and pedestrians in all directions.

Everyone seemed in a hurry except a sixteen-stone patrolman, who moved along at a majestic pace. He stopped, twirled his truncheon and looked Kerr up and down, as though waiting for him to give an account of himself.

"Could you tell me the way to Irishtown, please?" asked Kerr, timidly.

"Turn right at the end of the block an' folly yer nose," he replied. "But say lad, you're headin' for a tough quarter. If you've any money, stow it away good; or some slicker'll have it before you're much older."

The further Kerr walked, the less he liked the look of his surroundings. The long stone-paved street was lined with tenements which soared three and four storeys above dungeon-like basements. The blue strip of sky, showing between the roofs, served to emphasise rather than dispel the atmosphere of gloom and despair which pervaded the district like a blight. Washing hung out to dry on make-shift poles and lines, pitiful pennants of poverty and distress. Dirty emaciated children swarmed on the side-walks, climbed the lamp-posts, squirmed through railings, or leaned from high windows in deadly peril. They upset ash cans in search of scraps of food, banged the flagstones with empty cans and chattered like starlings. Here and there an old man or woman stood in a doorway, surveying the scene with lack-lustre eyes and an air of resigned indifference. Slovenly women shuffled on errands or stood gossiping with neighbours. Tired and hungry, Kerr spotted an eating-house and walked in.

The interior reeked with the smell of rancid grease, walls and ceiling were mottled by countless generations of flies. A sallow-faced slip of a girl leaned against the table behind which Kerr had squeezed into a seat.

"Pork an' beans ten cents," she announced dully, eyeing his bags with concealed interest.

"That'll do me," replied Kerr amiably, relaxing his tired limbs. He ate without relish and enquired about accommodation.

"'Fraid we can't put you up — but I'll ask Maw," the girl replied, bestowing a smile as he paid for the meal.

"Bad cess to them flies," interrupted a voice, as a blowzy woman of fifty or so stepped heavily into the room.

"He's from Ireland, Maw. Lookin' for a place to stop."

Mrs. McGarry glanced from one to the other.

"Can't make out what brings you lads over here at all," she said wearily. "Divil a much there's to be got as far as I can see. Why for every one that makes out well in th' States, there's a thousand that goes to the bad, and ne'r a word about them. Hate to refuse ye, laddie, but there isn't room for a mouse in th' house."

"What about Lembergs?" the girl suggested.

"It ain't right sendin' a decent gosoon like him to a lodgin' house, Sadie, but he'll be lucky if he can get put up even there."

Kerr thanked them, gathered his bags and went into the street.

"At Dunphy's corner, over th' saloon; if ye don't get fixed up, come back," Sadie shouted after him.

The sign "Good Beds" over the door, indicated his destination. Kerr went upstairs and enquired at the three by two booking office on the landing.

"Kom see de beds," said Dutchy, peering at the newcomer through horn-rimmed glasses. "Dwelve zents a night," he continued, leading the way through a bare, gloomy corridor to a windowless room at the end. He struck a match and lit the gas. A double row of bunks filled with straw, sacking and worn-out clothing, lined the walls.

"Dis one," said Dutchy, indicating number six. "Youse lucky, jus' one lef'."

Kerr's heart sank. The smell of the place made him feel sick. "Have to make the best of it for to-night anyway," he thought, throwing his bags into the bunk.

"Rums open nine at night, ol out at nine in de mawning," said the proprietor, closing the slide of the pay-box.

Kerr strolled around the neighbourhood, trying to get his bearings. At nine o'clock he returned to the lodging house, footsore and weary.

The fish-tail gas jet gave just enough light to see dimly in the room. He spread a few clothes in the bunk by way of a mattress; his carpet bag did for a pillow. A huge muscular man, stripped to the buff, climbed into number five. Kerr took off his boots, hove himself into bed and fell asleep before he had time to think any more.

He awoke with a start, trying in vain to recall his whereabouts in the darkness. The heat was intense. Then he realised it was no nightmare, he was gasping for breath in an atmosphere thick with the stench of filthy humanity. Vermin swarmed around his ankles and skimmed over his body, quicker than his brain could follow their movements.

His first impulse was to get up and run out of the place, but the thought of the few dollars sewn in his belt caused him to hesitate and change his mind. The occupants of the other bunks moaned and scratched between snores. Tough customers, if they were disturbed by an unusual noise, some one would surely start a rough house and he'd probably lose everything and get beaten up into the bargain.

A distant clock chimed two, Kerr tossed about in acute unease and heard it strike three, four and five. At six in the morning the attacks of the vermin eased off and they retired, bloated like ticks, to their cracks and crannies for the day.

On his way out, Kerr asked Dutchy to store his bags. He could not face the thought of stopping there another night. Breathing deeply of the fresh air, he strolled around to get acquainted with the town.

A corner boy, on the lookout for a softie, cadged a cigarette.

"See ya don't know yer way around Bud," he remarked. "It's simps. Avenoo's run no'th an' sa'th, streets cross 'em east an' west. A blind man could find 'is way aroun' Noo York 'cept below First Street, where th' roads run haywire." His voice suddenly dropped to a confidential whisper. "Any chance of a dime, Bud?"

Sixth Avenue was a drab thoroughfare, grimy from the smoke of elevated trains that thundered over the west side-walk every few minutes. An uneven line of dingy shops and restaurants, pool rooms and sinister-looking dives, huddled in the shadow of the platform as if shunning the honest light of day.

The milling multi-coloured crowd fascinated Kerr. Workmen hurrying about their business with tool bags, porters with baskets on their heads, carriers pushing hand-carts through the traffic, women elbowing their way from shop to shop; unemployed and unemployable obstructing the busy folk, pestering everyone for the price of a meal.

The general air of poverty and distress was relieved at times by snatches of laughter and song, as a saloon door opened wide to a bright and warm interior; an irresistible temptation to unfortunates outside, to walk in and forget their troubles.

At the junction of Sixth and Broadway horse trams on iron-shod wheels rattled to a stop to let down passengers. After a pause they started again, driven at an unmerciful pace by whip-cracking buckaroo drivers.

Kerr crossed over and turned up a side street into the comparative quiet of a tree-lined residential quarter. Unexpectedly he emerged into the magnificent environment of Fifth Avenue and paused for a moment to take in the scene.

A mansion, half a block long, with flower-decked windows and massive doorways was visible through a railing. Periwigged flunkeys in red plush uniforms fussed around a carriage.

"Show place o' th' town," remarked a seedy, shabbily-dressed youth, who, like himself had stopped for a peep at the scene.

"Who lives there, the President?" asked Kerr, in an awed tone.

"Nope, it's the Astors' place; but that's nuthin', friend, they've another in th' next block, in case y' missed seein' this one."

"Swell place!" remarked Kerr, admiration in his tone.

"Naw kid, y'don't get me thinkin' that way," returned the youth sourly.

"Why, don't you think it's wonderful to be a millionaire an' have a place like that?" replied Kerr, a bit puzzled.

"W'en I know there's thousands o' families in Noo York, without a roof over their head, an' see one family with two empty palaces, it gives me th' sick every time I look at 'em. Th' sharks that grab all th' money get all the praise and admiration too," he continued, "even from th' poor slaves that sweat an' starve to pile it up for 'em. I reckon the whole god-dam lot should be blown up."

"Here, have a smoke," said Kerr, suddenly taking a liking to his acquaintance for his candid outburst.

"Thanks. My name's Crane. Steve to you," he replied. They lit up and strolled towards the Park.

The Avenue was on afternoon parade. Elegant ladies went about their shopping, daintily holding their long skirts clear of the dust. Dandies in top hats, frock coats and patent leathers stepped out on social calls. Cigar smoke trailed in the air.

Even the traffic seemed muted and orderly on this highway of fashion. Society belles, vaudeville stars, double-chinned dowagers, dyspeptic millionaires and wealthy playboys whirled along in carriages with attendant footmen in gorgeous liveries.

"I'm caretaker of the Arts Club in th' Bowery; get the afternoons off," said Steve as they sat on a bench in Central Park.

Kerr told him his story. His companion listened sympathetically.

"Sorry I can't put y' up, Bud. There's thousands like yourself, without a bed, an' don't know where the next meal is comin' from. Wish I could do somethin.' But say, friend, we've a Debatin' Society that meets every Saturday. Come an' get acquainted. Some guy may put y' wise to a job."

Kerr thanked him, glad to have made a friend.

"If I ever get down your way I'll look you up," he promised, as Steve got up and left for the Bowery.

Kerr dozed on the seat. When he awoke it was sundown. After a short stroll and a snack, he returned to look for a bench on which to pass the night. There was space on one already occupied by two scarecrow men and a poor old woman. He sat down. No one spoke. As darkness fell, an army of slum-dwellers and tramps converged on the park and took possession. Whole families spread their bedding on the grass, to rest under the stars. For the first hour or so, Kerr felt warm and comfortable, but after that, sitting on the hard bench became wearisome.

As the clocks struck the hours away, the benchers nodded and dozed, but were denied the comfort of sleep by demons of cramp and cold, which caused them to shift and shiver as the night wore on, yet those discomforts were tolerable compared to the torment of staying in the choking bug-infested tenements.

At day-break Kerr had to get up and walk around several times to restore circulation, his personal distress curiously eased by the thought of the wretched multitude around him. If the people suffered from cold and exposure in the summer months, how could they hope to survive a winter, he pondered. He pulled down his cap and buttoned up the collar of his coat as a piercing breeze began to blow in from the Hudson River.

That night dispelled his boyish dreams of fame and fortune once and for all. The thought that when his small store of money was spent he would be in the same straits as the host of unfortunates who were condemned to die from exposure and want, goaded him to action.

He tramped the streets, rode the El. and jumped on the crowded horse cars in a desperate search for employment, gradually acquiring a sense of direction and a knowledge of the main streets and avenues which intersected the island of Manhattan.

Kerr was amazed at the number of educated folk who were dragging out an existence on the bread-line. There were immigrants who couldn't settle down, disbanded soldiers and sailors, natural-born criminals and tramps, mental and physical wrecks and broken-down professionals who had worked and lived with the highest in the land.

Drink, drugs and gambling had brought ruin to many, some were stricken by ill-health; others were the victims of accidents; more again seemed to have been dogged by misfortune all their lives. Most of those he got into conversation with had a hard-luck story to tell, usually as a prelude to begging the price of a meal.

Now Kerr was able to understand why New York was the most overcrowded city in the world. The influx started when the Civil War ended and ex-soldiers returned to resume their normal way of life. During the same period, poverty and unemployment in the great cities of the Old World forced millions of people to emigrate. North America, with its boundless space and unlimited opportunities, seemed to hold prospects no other land could offer.

Sailing ships were fading from the sea in favour of quicker and more reliable steam boats. Shipping magnates foresaw a demand for passenger accommodation that would last indefinitely. Great shipping companies were founded. Yards in Glasgow, Belfast and the Tyne worked day and night to produce bigger and faster ships.

In the great seaports of Europe and Asia, hordes of immigrants queued up for transport to New York. The United States had the name of being a free country where one could build a home and raise a family at a decent standard of living, provided one abided by the Law.

When the great influx gathered force the Government imposed a head tax to reduce the number of immigrants, but this failed to stem the flow. As time went on and the vast sums accruing to the State from immigration became known, vote-catching politicians conveniently switched their outlook and changed their party cries to "More heads, more dollars. Why worry? Let 'em all come."

From that time on scarcely a day passed but some vessel arrived from a foreign port and dumped its human cargo on the already overcrowded shores of Manhattan. Building sites became as valuable as gold mines and property owners became millionaires overnight. Contractors and rackrenters combined to exploit the situation, unembarrassed by the law. As ground was beyond price, houses were built up on existing foundations until the walls soared eight, nine or ten storeys above the pavement. They were set so close together that sunlight was shut out from the streets below. Existing buildings were divided into tiny rooms, many without windows for light or ventilation. A water closet and tap on each landing served the hapless tenants.

Prosperous American families, surrounded by the rush of foreigners and ever-encroaching slums, sold their homes to speculators who at once converted them into tenements. In 1880, one and a quarter million of the inhabitants of Manhattan were forced to live in so-called apartment houses. Exorbitant rents were charged for the rooms; families had to slave day and night and do without the necessities of life to pay the weekly rent. If through illness or unemployment they were unable to pay up, they were thrown out by bouncers, to exist as best they could in rat-infested alleys or damp underground cellars, the last resort of down-and-outs before death put an end to their sufferings.

Irish people settled on the west side of Broadway. Negroes went to Harlem. Italians took over the Bowery. Poles, Finns and Greeks formed colonies along the bank of the East river. Turks, Lascars and other orientals swarmed along the southern tip of the island. Jews wedged themselves in between the boundaries of the main groups. Each settlement was a little world unto itself where the people spoke their own language and fostered the traditions of their native land.

Broadway, running like a backbone through the middle of Manhattan, was reckoned the respectable, business part of New York. The streets branching off on either side spread out into a maze of lanes and alleys which extended for miles. Here, in crumbling tenements, tin-roofed shanties and every conceivable type of makeshift habitation, dwelt the millions that were building the world's greatest city. With three million inhabitants engaged in an elemental struggle for existence, New York was a place where the chance of living to a ripe old age was slight. The vast slum-lands lying east and west of Broadway were never cleaned. Slops and garbage were thrown from the windows; a stench from accumulated filth filled the air. Flies carried germs from one plague spot to another, and epidemics raged. Over three hundred and thirty thousand people struggled to live in each square mile. There was scarcely room to breathe and the mortality from tuberculosis was so high that sections of tenements bore the sinister name of "lung blocks."

As often happens where jungle conditions prevail, the savage, lawless elements fought their way to the top and enriched themselves at the expense of those who strove to live by honest means. Gangs extorted tribute from shopkeepers and others engaged in legitimate business; the victims knew from experience that it was better to suffer in silence than go to the police.

At the time New York was governed by a Society known as Tammany Hall, which derived its name from the building where the society was convened in 1789.

The founders, inspired by the altruism of Tammany, a renowned Indian sage, were men of wealth and culture, animated by a true civic spirit. From 1800 onward, most of the mayors or bosses in New York were nominated by Tammany. Under its guidance a sprawling collection of towns united and became a vast city which covered the island of Manhattan.

For administrative purposes New York was divided into districts and wards, each controlled by a leader, nominated by Tammany. The leader organised social functions ostensibly to help the unemployed, and sponsored Church collections, charities and other meritorious works so as to gain the confidence and the votes of the masses at election times.

About the middle of the century, gangster elements infiltrated the society and finally gained control. With the judiciary and police force under its thumb, Tammany represented supreme authority, High Society, vested interests, and above all, security, at a time when millions were yoked to a life of industrial slavery and the voice of anarchy was rising in the slums.

With fabulous wealth on one side and grinding poverty and starvation on the other, many who could not obtain an honest living were forced to co-operate and lead a life of crime.

Tammany leaders, unable to enforce the Law in congested districts where the people were as vicious as scorpions, allied themselves with their contemporaries in the underworld and became shareholders in a billion-dollar business of organised crime. New York became a city of unrestricted sin and corruption. Police collected taxes from thousands of gambling joints, brothels and drinking dens. Traders could sell underweight coals and groceries, watered milk, substitute butter and unwholesome food, without fear of arrest, so long as they bribed the cops on their beat. Anyone who dared to complain was beaten up and often left for dead.

No one was safe; even the world-famous inventor Thomas A. Edison was victimised by the racketeers. After years of frustration he had obtained permission to build an electric power station and install a public street lighting system in New York. He was out with his assistant one morning in Pearl Street setting tubes in trenches for encasing the main cables, when a Tammany Hall official touched him on the shoulder and said, "Say, Mr. Edison, this is dangerous work you are doing. It will have to be properly supervised or stopped altogether. The Department of Public Works requires that you shall have five inspectors to check this work and their salaries will be five dollars a day, payable at the end of each week."

With fourteen miles of cable to lay, the inventor, sickened by the thought of being further harassed and delayed, agreed to the impost at once. The cable-laying was allowed to proceed. The inspectors turned up only on Saturday mornings to draw their pay.

Harassed by rack-renters, usurers and racketeers, thousands of decent working-class families were forced to live in permanent poverty and demoralising slums. Time and again social reformers arose and attempted to overthrow the unholy trinity of millionaires, mobsters and uniformed thugs, who made a mockery of justice and fattened on the proceeds of vice and crime, but as the cynics said, "They're only hittin' their heads against a stone wall."

It was impossible for the immigrants to weld themselves into an effective political force owing to the confusion of tongues. To make matters worse, party leaders spent money freely at election times. More often than not the poor man's only weapon, the vote, was sold for a drink.

4

ONE BLAZING midsummer day, Kerr wandered into Baxter Street, an outpost of Jew-town. The side-walk was lined with rickety stalls on which secondhand clothes, fish and vegetables, scrawny chickens, goat's meat and long twisted rolls of bread, were set out for sale. Flies circulated in clouds, hawkers chanted, sallow-faced women and children milled between the stalls, ankle deep in filth and rotting vegetation. From open windows on either side of the street, came a continuous rumble from hundreds of sewing machines, for this was the centre of the world's largest clothing industry and every tenement was a factory in production. Pushcarts were trundled from house to house, piled like haystacks with garments in every stage of manufacture.

A man steering a handcart cursed loudly and long as it tilted up and swung him off his feet, but his mate, supposed to be pushing behind, was too weak and apathetic to be goaded into action by mere words. The barrow-man pulled over to the sidewalk, steadied his load against the kerb and said:

"'Ere, Ikey, 'ad abaht enough o' you an' yer 'angin' on behind. 'Ere's wot's due to yah. Sling yer bloody 'ook."

"Any chance of a job mister?" Kerr asked on the spur of the moment. The Cockney looked up wearily.

"Orl right, give yer a trial mate. But y'gotta work an' be snappy. 'Ump this 'ere load up ter Levi's on th' third floor fer a start. I'll watch th' barrer."

Kerr heaved a bale on his back and climbed the dark rickety stairs. As he went up, the combined smells of cooking, cloth dyes and cats made every breath a penance. Old Levi framed himself in the doorway on the landing.

"You vos late again," said Levi, hands shoulder high, palms outward in protest. "Stack 'em up 'ere." He pointed to a small space near the door jamb.

For lack of space Kerr had to build the bales ceiling high.

It was backbreaking toil, but after overcoming his first feeling of nausea, he felt thankful to have a job, even if it was worth only a quarter-dollar a day.

The pair worked from seven in the morning to late in the evening, in a district bounded by Baxter, Bayard and Hester Streets, where as many as two thousand five hundred Jewish people were packed into a single block. In every house there was the same congestion, from dungeon-like basement to eighth-storey attic. Every room was piled high with bales of cut cloth waiting to be sewn; many old floors and walls were cracked and dangerous from overloading.

Except for two or three sewing machines and a small table or shelf to press the finished garments on, there was no furniture. Bales were stacked to form tables, chairs and beds. In a small clearing, the owner of the sweat-shop, his family and employees worked at a feverish pace, for ten to fifteen hours a day, basting, lining, sewing button holes, or pressing the finished garments in a cloud of steam, and not daring to speak for fear of losing time.

The sweat-shop system forced employees to work harder for less money. Many a prosperous sweat-shop proprietor had begun by starving himself and his family to save a few cents a week. When he put by enough money to rent a room and a couple of sewing machines and learned enough English to negotiate with a factory for supplies of cloth, success was in his grasp — for thousands of skilled tailors were walking the streets clamouring for work, glad to be taken on even as "learners" and willing to accept whatever rate of pay the sweater chose to give. The sweater increased his profits by reducing rates of pay still further and was soon in a position to rent another room, hire more sewing machines and start another sweat-shop.

Children of the ghettos struggled for survival under shocking conditions of overcrowding and neglect. Stunted from lack of air and sunlight, growing up half naked and half starved, they were set to work as soon as they could be taught to thread a needle or knot yarns of cotton, usually at the age of eighteen months. At three they were skilled at sewing on buttons, or finishing button-holes and their labour was worth up to thirty cents a week.

Only the strongest survived; the sickly weakened and died, but there were plenty more on the way to take their places. A great proportion of the youngsters who lived, developed horrible deformities and loathsome diseases and at the age of five or six had the puckered, wizened faces of old men and women.

Kerr struggled on gamely, well aware that in the desperate conditions that prevailed, he was fortunate to get employment of any kind. He slept at night in tenement hallways, or, if the weather was fine dozed on a doorstep until it was time to eat and go to work again.

No matter how hard he worked, the Cockney was never satisfied.

"D'you think I'm a slave?" Kerr blurted out one day after working for eight hours without a break for a meal.

"Wot's narkin' y' mate?" retorted Alf. "If we don't work like this, someone else'll get in on our beat an' our clients'll give us th' go-by." He went on to explain that things had gone against him since he came to New York and that he was barely making a living with his "barrer." When Kerr discovered that Alf was little better off than himself, his rancour changed to sympathy for a brother in distress and after that they worked in harmony.

One evening, after finishing work he sat down on a doorstep, tired and depressed. It seemed clear that if he held on to his job at one and a half dollars a week, he could never hope to attain a decent standard of living, or even save up enough money to buy proper clothing for the winter. Although most of his earnings went on food, he was as thin as a whippet and sometimes felt so done up in the morning that he wouldn't have gone to work at all but for a sense of loyalty to Alf.

On Saturday, the Jewish Sabbath, there was no call for transport. The pair met in a café to reckon up accounts, plan work for the following week and relax.

"Abaht that dollar due from ol' Jacobs," said Alf querulously.

"Guy in the fifth front, Hester Street? — said he didn't owe a thing," replied Kerr.

"Lousy shonk," retorted Alf, spitting. "No mo' bleedin' credit for 'im."

"I could hardly get the stuff into the room with the crowd that's in it anyway," continued Kerr.

"There was old Edzel, his missus and five kids and the three learners who haven't a word of English between them. None of them would get out of the way, never mind help."

"Just like 'em," mused Alf.

"Old Edzel was pressing pants," continued Kerr, "his missus was cooking the dinner and the three learners were working the machines like engines. The girls were all black from the cloth dye and the baby was threading needles like magic. Thought the poor kid was a monkey at first."

"They get it tough," Alf conceded, "but ater all, th' Jews' motter is every man for hisself, so wot can yer expect?"

"Wouldn't it be better for everyone if they built factories?" said Kerr naïvely.

"Facktries!" snapped Alf. "Oo's gointa build facktries w'en there's people makin' millions outa things as they are? W'y, it's all a set-up, kid. Th' sweaters makes the clothes for 'arf nuthin', th' big shots sell 'em in th' stores at a thaasan' per cent profit. Wot they can't get rid of is sent to 'and-me-down shops all over th' world. No rates, no over-'eads, no competition, w'y its all profit; so them as could ain't goin' ter try an' change things."

Kerr began to see the logic in Alf's reasoning.

"W'ile I 'old that th' Jews isresponsible for this 'ere 'ell upon earth, it don' mean there's no one else ter blame," he continued. "Look at yer 'igh Society on Fifth Avenoo, them as owns these 'uman pig-styes an' th' graand they stands on. Christians! Goes ter church an' all that." He spat violently.

"It doesn't seem right that some should have everything and others have to go without food," said Kerr.

"D'ye fink them 'igh ups as spouts Christianity gives a hoot if th' rest o' th' world dies o' starve-ation? Not them. Gamblin' on th' Stock Exchange, cornerin' food, squeezin' every ounce they can get outa th' poor; with th' Guverment an' perlice behind 'em ready ter smash daan anyone that cuts up rough an' starts a riot.

"W'en I sees wot them church-goin' 'ippercrites closes their eyes to, I begins ter fink there ain't no God, cos if there was 'E wouldn't stand for it! "

Ending on this bitter note they parted, Alf for his doss-house and Kerr to a tenement hallway, to flop for the night.

A fortnight later, Kerr had a bad turn. Every bone and muscle ached and he shivered with a deadly chill although the day was bright and warm. His main worry was that he couldn't get word to Alf. Hardly able to drag himself along, he groped his way to a saloon and had a few tots of rye whiskey. The raw spirit burned his throat and made his head spin, but it checked the fever and after a couple of days he felt well enough to go back to work again.

His luck was out. Alf had started another man and Kerr had no option but to begin all over again tramp-ing the pavements in search of another job.

He wandered into the outcrop of villages and settle-ments along the east side of Broadway, got completely lost and was mis-directed so many times by people who had only a little knowledge of English, that he be-came exhausted and could go no further. For days he tramped through the Spanish, Bohemian, Russian and Scandinavian settlements and finally came to a vast dreary waste-land, where the city garbage was dump-ed by an endless train of waggons until it overflowed into the lanes and alleys of notorious Mulberry Bend.

Here, in gaunt derelict houses an army of men, women and children of the rag-picker class, eked out an existence, like a horde of rats in a rubbish heap. These poor creatures were too weak and apathetic to engage in the normal struggle for a livelihood. The sight of them inspired Kerr to push forward and away from the place, lest an insidious torpor should come over him and undermine his will to carry on.

5

ONE MORNING he walked foot-sore into Barclay Street, Greenwich. It was Sunday. Bells rang and crowds were on their way to eleven o'clock Mass at the church of St. Peter. Some arrived in glossy carriages, but most were of the working class and bore the stamp of poverty. The older women had shawls on their heads, and the children were scantily dressed, some of them barefoot. Youths and girls wore cheap, ready-made clothes just like the crowd Kerr had seen gather outside the church in Borris-in-Ossory, only a few short months before.

He walked up the centre aisle, thankful to be among his own folk again. When the choir sang familiar hymns, thoughts of home crowded into his mind and his eyes filled with tears. The priest delivered a sermon on the evil of a materialistic outlook and exhorted his flock to remember the words of St. Paul; "Whatsoever doth it profit a man if he gain the whole world, if he lose his immortal soul?"

After Benediction, the congregation streamed out into the street and many stopped to chat in groups along the side-walks. Kerr joined a crowd at the corner of the street that jollied the passing girls. He told his tale and found they were eager to help a gos-oon from the old country.

"Might get fixed up at the docks tomorrow if yer luck's in," said Mick Hanlon, a sallow-faced, stock-ily built man. "But it's hard goin'," he continued, holding out grimy, calloused hands as proof. With only a dollar and twenty cents between him and the bread-line, Kerr wasn't choosey. He thanked them all and bid them "Slean libh," as Mick's girl friend came along.

Kerr passed the night in the hall of a Pike Street tenement, near the place he was to meet Hanlon in the morning. He was sheltering in the doorway of a dockside café, when his friend appeared in the gloom.

It was bitterly cold — too cold for talk. Fog horns wailed on the East river, winches rattled on wraith-like ships, warehouses loomed up suddenly, as they picked their way through the mist. A stevedore's whistle shrilled in the distance, and clogs clattered on the cobbles, as the crowd made a last minute rush to beat the clock and the fog.

A huge man stood at a table set between the dock gates. He was calling out names from a list. Knots of men stood on either side at a respectful distance.

"Hanlon," he shouted.

"Folly me an' chance your arm," said Mick.

"This is my mate," he said, passing the table.

"Vell, vot for you vait?" roared the giant, purpling up, as he thrust a tally into Kerr's hand. "Lose dat an' der's no pay," he warned.

Kerr grabbed the tally, skipped out of range and sprin-ted along the quayside to where the full-rigged ship 'Thor' of Copenhagen, lay moored at the coal wharf.

They climbed the dim-lit gangway to the after-deck, scrambled through a chaos of ropes and hatch covers and down an iron ladder, to the bottom of number two hold.

"Can't make out how I got taken on with that crowd waiting," said Kerr, elated at his success.

"Shhh, mind whatya say," returned his friend. "Those guys are on th' black list. Wouldn't dib out money for the bosses' beer."

"How's that?" enquired Kerr.

"You're supposed to get a dollar a day at this racket, but you only get seventy-five cents. Th' ganger keeps the other twenty-five," answered Mick.

"But that isn't fair," retorted Kerr.

"Listen, Bud," replied his friend. "Take things as you find 'em here an' don't start kickin'. It don't pay. There's a crowd here on the waterfront that 'ud kill anyone that tries to bust their racket."

The rungs of the ladder sounded dulcimer notes, as two more men descended into the hold and sat down on the bilge boards for a smoke.

"Been at this graft before?" said a Glasgow voice within range of Kerr.

"Don't answer," said Mick in a guarded tone. "You'll be my opposite number shovelling. Do as I do an' you'll be O.K."

A few minutes later there was a rumbling and scuffling as a chute was pushed into position overhead, then before a shout of "Stand clear below" had quite time to reach them, an avalanche of coals crashed into the hold.

"Grab that shovel an' get goin' for Christ's sake!" yelled Mick as Kerr stood, wondering what to do. He soon learned. The four men had to compete with one another, trimming coals from the centre to the far corners of the hold.

The non-stop flow of coals from above and the ever-rising heap under the chute were the pace-makers. There was no let-up. Each man had to lift the big lumps and run with them to his corner, then return to the heap and shovel like a madman, to keep his section level with the others.

The fog above deck cleared at day-light and sunbeams filtered through the dust in the hold. The bulky figure of Krupp appeared at the yawn of the hatch.

"C'm on, youse tired bawsters," his voice boomed through the hold. From that time he came every half-hour or so to worry the men with a brutal obscene tirade that carried a threat in every sentence.

At ten o' clock the hold was like an oven and the dust so thick that breathing was difficult. It was the longest morning Kerr had ever put in. His hands were blistered and bleeding and his whole body pickled with sweat and grime.

When at last the whistle sounded for dinner break, Kerr joined in a rush to an eating house and collapsed on a form.

"Here, Bud, get this down an' you'll be all right," said his comrade, handing him a bowl of stew. When the whistle to start work sounded, Kerr felt game enough for the next round, in what he now knew to be his fight for survival.

The afternoon dragged on. Krupp got on everyone's nerves, by sneaking up to the hatch and bawling out anyone who dared to stop for a breather or even for a drink of water. When at last, the break-off whistle sounded, Kerr went wearily down the gangway and trudged through the streets and alleys, until he came to a block of tenements. He staggered like a drunk up the steps into the hall of the first one, curled himself up on the flags, sick and sore in body and mind, and slept in a state of torpor until dawn.

After three days, during which his tortured muscles almost refused to function at times, Kerr felt he would be able to hold his own. He formed the German habit of carrying some bread in his pocket and munching a bit when nearly overcome by fatigue.

Scotty, his opposite number in the hold, was a skilled man who had worked at some of the great ports in Britain.

"Thought Clydebank was tough, but it's nothin' to this land o'Liberty," he remarked to Kerr, as they walked home one evening. A paroxysm of coughing racked him and he spat red on the pavement.

"Can't you get anything for that cough?" said Kerr, shocked at the sight.

"Not a thing, Bud, it's th' dust that causes it. Gets everyone in th' coal racket, in a few years."

"Why don' you look out for another job?" enquired Kerr.

"Guess it's too late now. Can't sleep at night thinkin' o' what'll happen to th' wife an' kids if I stop," said his mate grimly.

Kerr turned into Munroe Street, a narrow canyon between two rows of tenements. Poor brave Scotty, who neither smoked nor drank. All he lived for was his wife and five children, cooped up, in a twelve by ten box-room on the fifth floor of a Cherry Street tenement, his home. When Scotty died, his whole family would starve to death as a matter of course. Terrible to think of it, but then, he reflected, the same doom awaited millions of people who lived in the shadows of want and despair in the heartless city of New York.

Half-way down the street, Mick was waiting for him on the front step of the dosshouse, where they shared floor space. Kerr sat down beside him, tired and dejected. A tide of people coming home from work flowed past, filling the air with a Babel of sound.

"Ever set eyes on such a mob?" said Kerr in amazement.

"They're from every part o' th' world an' there aint two of 'em the same colour," replied Mick.

"There's blacks an' whites, Injuns, halfbreeds, China's an' Japans. Jasus, most of 'em couldn't tell where they're from," he opined.

Two pretty, smartly-dressed girls came out of the hall behind them, giggling as they passed with skirts held close as if to avoid contamination. Kerr flushed with anger but checked a hasty retort.

"Arrah don't mind th' haybags," said Mick so all the street could hear. "It's the likes o' them that 'ud look down on th' likes of us anyway," he continued, spitting on the pavement.

"Shocking to see kids like that going on the streets all the same," replied Kerr moodily.

"That kind wouldn't work when there's easy money goin'. Why half the girls in town is on the batter," concluded Mick, bitterly.

"Here, isn't it time we started looking for another job?" said Kerr suddenly. "Let's get cleaned up and go down the waterfront."

6

HALF AN hour later they were on South Street embankment. An incandescent blaze of lights from saloons, poolrooms and funfairs stretched away into the distance, twinkling like beacons against an inky sky. Pools of light shimmered in the waters of the East River.

"Let's go in an' see how things are goin'," said Mick, pushing through the doors of the Nantucket Bar, the biggest saloon in dockland.

"Aw, c'mon, man, what are ya havin'?" said Mick irritably, as they stood at the vast horse-shoe bar. Kerr hesitated, rembering all the warnings he had had about drink from the old folk, but, when the bartender put down two schooners of beer on the frothy zinc counter, he couldn't back out before the crowd.

The place was packed, the atmospjere hazy from the smoke of tobacco. Men sprawled against the counters engrossed in boozy arguments.

The wall seats were crammed with old women and girls in various stages of intoxication, who joined in the chaos of maudlin songs. Clustered around, were pathetic groups of dirty, emaciated children, who bawled continuously until their hunger was relieved with packets of peanuts or popcorn.

Amid the general uproar and occasional crash of glass, tough-looking bartenders in shirtsleeves and dirty aprons weaved among the tables, skilfully serving drinks, collecting empties and keeping an eye out for chancers or other potential trouble makers.

After the third round of drinks, Kerr began to raise his voice and talk silly.

"Go steady," said Mick in a fierce whisper. "See who's over there."

Kerr followed his glance and there at the opposite side of the bar with his elbow on the counter was Krupp, boss stevedore of the South Street waterfront. The end of a big cigar was stuck between his yellow buck teeth. His big square head and bloated features, bore traces of wounds acquired during his early days when he had fought for nickels in a Coney Island boxing booth.

A crowd of toughs and toadies milled around, trying to get on speaking terms with the boss, by calling for drinks out of turn and generally acting the maggot after the manner of their kind.

"Beats me how big bowsies like him get all the soft jobs and poor devils like you and I have to mullock and sweat, for every cent we get," said Kerr, feeling sorry for himself.

"It's all a system," said Mick in a guarded voice. "In every crowd o' men that goes working down a ship's hold, there's one flyboy who sets out to curry favour with the boss in the hope of getting a soft job later on. This bastard starts off by working faster than his mates and after a few months gets a few cents an hour more than the others as a pacemaker.

"After making a name for himself as a Company's man, he works all day at a killing pace and anyone who isn't able to keep up with him gets fired ('No use to us,' says the boss). For a few years he slaves like a machine, then he gets a permanent job as a ganger. All he has to do after that is to walk around with his hands in his pockets an' threaten to sack any poor devil that doesn't do three days work in one. That's th' whole secret, Bud," he concluded.

"Time to go," said Kerr, a bit sick of it all.

"Wait a minit, Bud, there's a guy over there, all het up an' makin' a noise, there's gointa be fun."

A few yards away, just opposite to where Krupp leaned on the counter, a crowd had gathered around a seedy, slightly built man with iron grey hair.

Putting down his glass of beer, he began to address them, in the manner of a true soap box orator.

"I'm here tonight," he announced, "to tell you that after a lot of hard work, obstruction and threats from you know who, we have at last got a number of our fellow-workers to join together and start the Longshoremens' Union of New York, with an office as headquarters right here in South Street."

"That's Abe Fox, th' Radical. Jasus, he must be mad to talk like that in a place like this," said Mick.

"We aim to get every dock worker on the waterfront into our Union," continued the speaker, "an' when we're all organised and of the one mind we'll set out our case, make our own terms an' talk to th' bosses in the same bloody language that they've been talkin' to us since we were kids."

"An' what if they don't agree to our demands?" enquired a timid voice.

"No more o' that slave talk," replied Abe. "We'll let 'em see what we're made of, an' that we're ready to go on strike an' fight for our rights if we're forced to."

"Good ole Abe!" yelled a voice that brought forth a chorus of excited applause.

"Wait a minit, hombre, what'll th' Union do fer us if we join it? Will it give us work?" someone interjected.

A burly man pushed roughly through the crowd.

"Guess youse is one o' those slick Alicks that gets th' mugs to pay in every week so's youse can make an easy livin' for yerself," said the intruder.

"C'mon, fellas, roll up an' give yer names in," continued Abe, not a bit put out. "If you haven't got the five cents entrance fee with you, it'll do next week. Don't be afraid to come up men. United we'll stand an' fight fer better times, divided we'll fall an' never raise our heads again. We've everything to gain an' nothin' to lose."

A hush fell as a swarthy, bull-necked man pushed forward.

"So youse is one o' them fightin' men, is ya?" he drawled. Suddenly, a bottle whizzed and knocked Abe's hat off, before splintering a mirror beside the clock. A fusilade of missiles followed, smashing tiers of bottles on the shelves. Lights went out. Men jumped over the bar and began to help themselves to the hard tack.

Women screamed and children shrieked as the men fought for the drink, flooring friend and foe alike with the same gusto. As the battle raged, bartenders, armed with big brass knuckledusters on each fist, lashed out in all directions, pulping every face that came within range until the floor was littered with knockouts.

Mick and Kerr, lucky to be near the door, were caught up in the stampede and landed sprawling on the pavement.

"C'mon, Bud, beat it before the cops come," said Mick when he recovered his breath. They sprinted through every short cut they knew until within easy reach of the dosshouse.

Next morning, on their way to work, they stopped at an open space near the corner of Catherine Street, where a large crowd had gathered in silence. They pushed through, to find out what was wrong and what they saw wouldn't do anyone good.

Lying side by side on a bloodstained patch of ground were the bodies of three men whose faces were battered out of shape. Two appeared to be dead and the third twitched and moaned in his last agony. A youngish woman knelt by his side saying incoherent prayers and repeating "Oh God, why wouldn't you listen to me, Abe, why wouldn't you listen to me?" between outbursts of tears. A few men and women knelt with bowed heads and someone recited the rosary.

Mick and Kerr had one glance, made the sign of the Cross and left. They joined the stream of dockers walking briskly to their work.

"That's what people get fer havin' anything to do with Unions," said a tough-looking man, loud enough for everyone to hear.

"Some o' th' bosses' work," another muttered.

"No one'll ever find out who did it neither," said another voice.

There was no more talk about Unions. Everyone put a guard on his tongue and those who knew anything, were afraid to come forward and help to being the murderers to justice.

7

WHEN THE 'Thor' finished loading, Mick heard that two ordinary seamen were required to complete her crew for the long voyage to San Francisco, but when he and Kerr spoke to the chief officer they were told that one man had already been picked. Mick got the only job that was left. The ship sailed three days later.

Ten years were to elapse before Kerr heard anything further about his mate. In the fall of 1898, he was reading a newspaper report about the American war in Cuba, and there, in a group photograph of men in naval uniform was his old buddy, Mick Murphy. It was the same Mick, who opened up the seacocks and sank the seven thousand ton collier 'Merrimac' in an attempt to bottle up the Spanish fleet in Santiago Bay.

The 'Merrimac' was manoeuvred into blockade position, while the guns of the fortress Morro Castle bombarded the ship from a distance of a few hundred yards. For his part in this heroic action, Mick and other members of the crew, were decorated by the President of the United States.

Once again Kerr found himself alone and friendless in his struggle for existence in New York. He slept in stinking tenement halls and lived a hand-to-mouth life on what he saved from his wages. Most of the dockers went on the booze on pay-day and were too sick from drink to turn up for work on a Monday morning.

When Kerr got wise to this, he came to the docks early and was fairly sure to get a couple of days' work at the beginning of each week.

In autumn, shipping became slack and work harder to get. The boss gangers took full advantage of the times and drove the men unmercifully to speed up loading and discharging ships. No one dared to hit back or even complain. They worked for the lowest rate of pay at a killing pace, until consumed with despair and broken in health, many died like beasts in harness.

In October, 1888, freight rates dropped to zero and many shipowners went out of business. The slump that followed almost paralysed the Port of New York. Thousands of ships lay at the wharves and buoys, a forest of masts idled against the sky.

The depression caused mass unemployment, on a scale that had never been known in the history of the Port. Crowds of starving dockers, incited by anarchists, began to loot and burn the warehouses from which they were locked out. A million men, women and children were reduced to the last limits of desperation.

Police and State troopers cordoned off trouble centres and beat the mob into dumb acceptance of its fate. As if to aggravate the general distress, grey skies and icy blasts heralded the approach of winter.

One morning, Kerr found himself in the centre of a mob that set out to ransack a general warehouse. As the doors were giving way under pressure of the crowd, a jangling fire brigade arrived and stopped at a near-by hydrant.

The firemen jumped to action stations, connected up hoses and turned them on the mob. Weakened by hunger and exposure and soaked with icy water, the rioters capitulated and fled. The riot squad, out for blood, beat up those who were unable to get away in time. Kerr got a savage kick in the back-side and ran a painful half-mile before escaping into the comparative quiet of Broadway.

It was late afternoon and snow was inches deep on the sidewalks. Kerr turned eastward at Houston Cross to avoid the bitter wind. Lights appeared in the shop windows, pedestrians cloaked their ears, wisps of steam arose from the dray horses as they clopped disconsolately through the icy slush.

Kerr walked on aimlessly, wondering what to do — where to go.

A dilapidated barrel organ tinkled out the opening bars of "Sweet Rosie O'Grady" but false notes sounded so often that the operator gave it up as a bad job, and trundled his machine over to a saloon to wheedle a few nickels from the boozers, with the more inspiring strains of "Knock him down, McClusky," the hit song of the season.

As daylight faded into night, the sleet turned to snow. Kerr never found it so hard to find a place to shelter. Suddenly he saw the words "Art Students' League" in paper letters on a high window. The words reminded him of his first acquaintance in New York, and his forgotten invitation to the "Club."

After climbing several flights of dark, echoing stairs, he crossed the landing to a door that showed splinters of light around the frame and went on into a large fusty room. The walls were lined with cheap oleographs and crayon and pencil sketches. A wisp of smoke in a large old-fashioned grate, seemed to emphasise rather than dispel the wintry squalor of the room. A candle, stuck in a bottle, flickered on the mantelpiece.

His footsteps sounded on bare boards, as Kerr walked through a laneway of chairs to a table at which a man was seated. Their surprise was mutual.

"Sit down friend and pull over to the fire," said Steve heartily, waving him to a chair. Steve was enthusiastic about the Club, its members and his career.

"Writing a book about the slums," he explained, gathering up some papers, "but writing's one thing and getting your work published another," he added.

At that time, neither knew that Steve was on the way up and was destined to become a celebrated author. His first book "Maggie," a tale of the slums, had a poor sale. The publisher lost money on it and would have nothing more to do with Steve. His next book, however, "The Red Badge of Courage," a tale about the Civil War, sold over a million copies and brought the author fame and fortune.

"But Buddy, what about yourself, how're you makin' out?" enquired Steve. When Kerr told him of his rough deal, Steve was silent for a while, then replied:

"Look here, Kerr, I've got by for a few winters in this city, an' reckon that anyone on the breadline at this time o' year is a gonner. From now on, we'll get snow for breakfast, dinner an' tea an' thousands of tramps'll fall down on the pavements, lie there an' never get up again.

"It's gruesome, Bud. Every day, squads of ash carts come out of the City depots, pick up the stiffs an' haul 'em over to th' Potters Field for burial. No post-mortems or burial services for them. Jest dump 'em inta ditches six feet deep an' cover 'em up. That's th' finish of 'em.

"Your'ne not lookin' too good Bud," he continued sympathetically. "I'd put you up here for the night, only I'd get the sack. As it is, I'm only drawin' enough pay to keep body an' soul together."

"Oh I'm O. K. Get fixed up somehow," returned Kerr, trying to look cheerful.

"But wait a minute Bud," Steve said. "Mebbe I can putya wise to sumptin', that'll getya outa this jam."

"What is it?" enquired Kerr eagerly.

"I read in th' papers last week that there's a new Secretary of the Navy appointed, by the name of Whitney. This guy's got sizeable ideas an' already he's talked Congress into building up a Navy that'll make the United States the greatest sea power in the world.

"They'll be spending millions o' bucks on ships and there'll be a boom on for years. There's a battleship ready for launching an' it's a cert they'll be looking for crews. Anyone'll get a job as a swab. Good clothes, a decent flop, four meals a day an' a number on Uncle Sam's payroll, what more does a guy like you want?"

Kerr had made up his mind before Steve finished. They walked down to the front door. Steve pressed a quarter dollar into his hand; Kerr took it under protest.

"Aw call it a loan, if ya feel like that," said Steve kindly as they parted.

After a hard night, dodging in and out of tenement halls between snatches of sleep, Kerr caught the Fulton ferry over to Brooklyn and set out for the Navy Yard.

He walked on cobble-stone roads, through a vast area of gas-works, foundries, rubber and chemical plants and other industries. A screen of smoke and acid fumes shut out the sky; the air was thick with nauseating stenches that made him long for the pure air and fragrance of the boreens in his native land.

After losing his way several times, he came to a district of tenements, flop-houses, saloons, eating houses and dens of thieves and on turning a corner saw a long railed enclosure, in the centre of which stood a signal mast with the Stars and Stripes fluttering at the truck.

The great straggling walls and outhouses of the naval barracks spread out towards the East River and the vast parade ground looked as dismal and forbidding as the exercise yard of a jail. A fed-up looking Marine, armed with a training musket, stood on guard at the entrance gates.

Kerr felt his spirits drop to zero, but he pulled himself together and struck up a conversation with the sentry. The Marine aimed a jet of tobacco juice at a stone and scored a hit.

"Might have a chance of gettin' signed on as a swab," he allowed. "Offis is over there at th' guardroom. Not puttin' y' off Bud, but I reckon you'd be just as well off in prison as joinin' up in that outfit," he remarked, gloomily resuming his beat.

Kerr crossed the parade ground and entered a long, narrow room. Four youths sat on a wall seat, talking in whispers. They moved to make room for him. On the wall behind were posters, setting out scales of pay and allowances for servicemen of all ranks. There were action pictures of sailors in bell-bottoms, stokers, ship's carpenters, stewards and petty officers, pinned to the walls.

On one wall were oleographs of commissioned ranks, dashing ensigns, dare-devil lieutenants, dignified captains and admirals in elegant gold-ribbed uniforms, their breasts decorated with rows of medals and ribbons for duty well done.

Looking up at them and reflecting on the power and glory they represented, Kerr felt very small indeed, but in the bare, chilly atmosphere of the room he derived some comfort from the thought that anyone could rise up from the ranks and become an admiral.

At the fifth "Next, please," he walked into the interview room and stood before an officer seated at a table. The Lieutenant asked questions about age and nationality, and a kindly-faced Sergeant of Marines took down the answers on a form.

"Period of service is for seven years," said the officer, "and as you haven't got a trade, you'll have to start as an ordinary seaman. 'Swabs' we call 'em here." He then went on to explain about rates of pay and allowances.

"You seem to be a bright sort of lad, so I'll give you a bit of advice," continued the lieutenant. "As a general rule, if you start as a deck rating, you'll have to remain one for your period of service, but if you show aptitude for a certain kind of work, you may apply for a transfer to one of the skilled branches of the Service. There's always plenty of scope for ambitious boys. It's really up to yourself, I wish you goodbye and good luck," he concluded.

After the interview, the Sergeant called Kerr aside and gave him some forms and tickets.

"Take these down to the Seamen's Institute off Sands Street and they'll fix you up with a bed and meals for the time being. In about a week, you'll get a note to attend the doctor. Look after yourself, laddie, all the best."

For over a week, Kerr enjoyed the luxury of a clean bed, reasonably good food and freedom from anxiety, in the spartan but kindly atmosphere of the Mission;

then came a letter with an appointment to see the doctor.

After a medical examination, the doctor asked him if he was fully aware of the responsibilities he was about to undertake.

"Will you swear to submit to the Rules and Regulations laid down for the government of Seamen in the United States Navy?"

Kerr replied "Yes, Sir."

"Then sign here, laddie. For the next seven years from this date, you'll be in the Service of Uncle Sam."

Next day he travelled with a crowd of new recruits to the Naval establishment at Newport, Rhode Island, to begin his training.

8

IT WAS late afternoon when they arrived.

"If I don't get a meal soon there'll be a showdown," said Nat, a fairhaired Scandinavian lad with a rebellious glint in his eye. Kerr liked his frank and fearless attitude.

Just then, the recruits filed into a messroom where plates were set on a long table. They stood to attention while the Duty Officer said the regulation Grace.

"For what you are about to receive thank God — carry on."

The ration of dry bread, a hunk of cheese and a mug of hot cocoa was frugal fare by any standard but mighty acceptable to hungry lads. When the meal was over the Master of Arms ordered them to fall in and gave them his formal introductory talk:

"Youse men remember, yer in th' Navy now and every order yez get hasta be carried out at th' run. Slackness or back-answers won't be tolerated. Discipline yerselves if yev any sense, an' keep outa trouble, but if ya don't learn to discipline yerselves, well, all I can say is Gawd help yez." With this piece of advice, he dismissed the class.

From that time on strict naval discipline was enforced. Everything was referred to as if they were living on board ship. The floor became the deck, the walls were called bulwarks and referred to as port or starboard instead of right and left, so the recruits were gradually initiated into the language of the sea.

Suddenly a boatswain's pipe shrilled and a raucous voice called out:

"Squad to line up at the Stores for issue of bedding."

Each recruit received a mattress, a pair of blankets, a canvas hammock and enough rope and cords to make slings and a hammock lashing. As they signed for the equipment, the hard-bitten pockmarked marine storekeeper said:

"Look after yer hammock, matey, it's the only bit o' comfort you'll have at sea an' if ya die you'll be sewn up in it nice an' slick, fer yer trip t' Davy Jones's Locker."

When all were served, the dormitory buzzed with activity as the lads got down to making slings and attaching them to the hammocks. Instructors showed them how to whip the ends of the cords or "nettles" and to splice eyes for slings at each end of the hammocks.

Learning to sling hammocks on hooks, fixed in beams high above deck level, was fun. There wasn't much room, only eighteen inches between the hooks, regulation spacing.

It was no easy feat for a greenhorn to get in or out of bed when the hammock swung away at every attempt, but the knack had to be learned; so had the art of making up and lashing a hammock for stowage.

It was dark when a clock struck nine and a boatswain's pipe shrilled:

"Watch below to sling hammocks and turn in."

After "Lights Out" the "Last Post" sounded, and faded away, impressive, lonesome and sad. The call seemed to have a solemn, personal message for everyone who heard it.

Although tired to the point of exhaustion, Kerr couldn't sleep. The sudden change from a jungle existence to comparative security and freedom from want caused a mental reaction. He eased himself gingerly between the taut canvas sides. The blankets and bedding smelled of tarry disinfectant but otherwise he felt clean, snug and contented.

He hadn't written home since his first week in New York ... must write home tomorrow for sure. Visions of his Mother, Uncle Michael, his brother Pat and little Mollie and Julia rose before his mind. Everything here seemed so foreign and strange. He became overwhelmed with grief and self-pity, pulled the blankets over his head and sobbed.

When he awoke, autumn sunlight streamed through the bare barrack windows. The air was nippy. He tried to guess the time and dozed off to sleep again.

Suddenly, a bugle blared the inspiring notes of the 'Reveille'. In a minute there was chaos. Two quartermasters swinging yardsticks went the rounds of the hammocks rousing sluggards with repeated whacks and shouts of "Wakee Wakee, lash up an' stow, lash up an' stow hammocks, come on there, show a leg."

Lads scrambled into trousers and jackets, searched for missing clothing and struggled with their task of making up hammocks, which was to become a routine chore for the remainder of their time in the Service.

In half an hour hammocks were stored in the racks and the decks cleared for scrubbing down. A Marine played a fire hose on the caulked planks and the recruits, barefoot, with trousers rolled above knees, began to scrub and holystone the deck. After the scrub-down a line of boys on hands and knees mopped up the icy water with cloths and squeezed them into buckets until the deck was clean and dry.

The work was hard, cold and uninspiring but it sharpened the appetite. While the mess tables were being set for breakfast, the "duty Peggy" was piped to the galley for rations of bread, coffee and beans.

A bitter snow-flecked wind whipped the waters of Narragansett Bay into whitecaps as the recruits mustered on the parade ground for first lesson in saluting and squad drill . . . and what a crowd of hoboes they looked in their varied headgear and tattered old clothes!

The portly, middle-aged Sergeant of Marines spoke to them in a gentle, fatherly sort of voice that somehow didn't seem in keeping with his grim, weather-beaten countenance and blotchy red nose. At the end of an hour's practice the wretched, shivering squad began to lose interest in the proceedings. When a recruit made a mistake some members of the squad began to titter and make rude noises.

A sudden change came over the big Sergeant. His eyes glinted and his nose flushed from its normal red to magenta.

"Sqad, stan' at ease. Stan' easy," he said in measured tones. "Seems to me there's sumpin wrong here, buckoes. We don't seemta unnerstan' one another. When I gets goin' with a new squad I allus remembers th' ole axiom. Y'can lead men if there's any good in 'em. Now yez see what happened when I tried. Well o' course there's allus them as won't be led an' hasta be dhruv.

"Youse is now gointa get a lesson in proper discipline an' if there's a geek outa any one o' yez, he'll be put in irons an' run in fer a stretch in th' hoosegow. Make a laugh outa yer poor ould instructor, would yez?

"Squad A-T-T-E-N-T-I-O-N, Right turn, D-O-U-B-L-E M-A-R-C-H" he roared and each time they got to the boundary of the vast parade ground, the order came from his leathern lungs, "A-B-O-U-T T-U-R-N, D-O-U-B-L-E M-A-R-C-H."

For a full hour he kept them at the double and when they began to flag and slacken pace, goaded them with every oath he knew.

When the bugle sounded for dinner the recruits limped over to the messroom, feeling that they had sold themselves into a life of slavery, little better than the darkies in the cottonfields around Uncle Tom's Cabin.

In the afternoon, the new hands marched over to the boat sheds by the jetty to learn the rudiments of seamanship. Standing in a circle around the instructor, they were shown how to make knots and splices.

"Must learnta make 'm so's y' can't make a mistake ... even in the dark," he told them. "Yer own life or one o' yer mates may depend on a knot. Might haveta lash yerself to a raft or a spar if th' ship's sinking, or mebbe you'd haveta send a man down from aloft when th' wind's howlin' mad an' th' vessel's rollin' rail an' rail in the sea."

The instructor's words carried the weight of personal experience so they heeded them and practised knots, splices, hitches and bends until their fingers became supple and sure. At two bells (five o'clock) the class was dismissed and all were free for the evening.

After supper most of the lads went over to the hammock room got out their hammocks and with the sailor's knack of improvising, doubled them over for seats. Great steel hammock racks, smoky ship's lanterns, spotless deck with pitch-black seams made a nautical picture with the boys sitting around, cracking jokes and skylarking.

It was a mixed, typically American crowd. Some had a touch of colour, a few had oriental features and there were three full-blooded negroes. From the general conversation Kerr gathered that most of them had joined up to escape death from starvation and thought themselves lucky to be fit enough to pass the doctor. As they got to know one another, they quickly formed into "Buddy" groups of two or three.

"Vell, vot d'ya make of it?" said a tall, lithe youth as he dropped his hammock and slumped down alongside Kerr. It was his acquaintance Nat, the only recruit who seemed to have a true vocation for the sea, or who knew anything about it.

"Aw, so so," replied Kerr. "Know anything about these parts?"

"Yea, I vos raised 'round th' coast here, know it fair well," returned Nat. Kerr found it hard to follow his quaint foreign accent at first, but encouraged his friend to talk.

"Mein fadder vos killed in a dory off Providence, run down in a fog. He vos fishing for clams. Bits of de boat vos vashed up on der beach nex' day but Dad vos never found. Mudder usta make nets, I usta help an' sell de nets. We live like dat fer years. Den machine nets come in and der is no vork for us. Ve vos ver' poor an' hungry, den Mudder die. I look, look fer vork, but der is no vork. Den I say I'll join Navy. Mebbe dis is bes' fer me."

Kerr was touched by Nat's simple and harrowing story. It was so true to life and typical of America.

He told Nat about his own home in Ireland and from that time on they were "buddies."

On Sunday morning everyone, except the cooks and stewards had a lie in. After breakfast all except the "duty" watch mustered on the parade ground. A staff captain, resplendent in full-dress uniform, took the parade. When inspection was over a bugle sounded and the men assembled in groups, according to their religious beliefs.

After dinner, Nat and Kerr strolled down to the water's edge. Fort Adams on a hill overlooking Narragansett Bay, dominated the entrance to the harbour. The fortress looked grim and invulnerable with batteries of guns peeping from embrasures at vantage points. The 'Constellation', an old frigate used as a training ship, lay moored off Coaster's Island.

They stopped at a formation of jagged teak piles which outlined the skeleton of what was once a long wharf. The sky was clear, the air crisp. Water lapped against the shell-crusted piles, in snappy, glittering wavelets.

"Dis vos a gread place in de ole days fer trade to Africa an' de Barbadoes," said Nat. "Ships usta load here wid rum from Newport, run over to de African coas' an' change der rum fer slaves. Den dey sail to Barbadoes wid der slaves an' sell 'em fer vork on de sugar plantations. Den dey buy a cargo of sugar an' sell at a gread profit in Newport, where dey make more rum from de sugar. Gread trade." He gazed sadly at the piles. "But now all finish," he concluded with a shrug and gesture of his hands.

One day, about a fortnight later the boatswain's mate piped:

"All new recruits to the Stores for issue of kit."

The squad cheered at this welcome break in their monotonous routine. Each lad was issued with a kit bag big enough to hide in. This was followed by a complete set of clothes and underwear for working in and another set for Sundays and Dress parades. None of them ever had so much personal baggage before. When everything was packed in the kit bags, it was as much as they could carry away.

They now wore jumpers and bell-bottomed trousers for the first time. When he glanced at his reflection in a mirror, Kerr felt anything but dignified in his uniform and little dicky hat. The uniform just didn't suit him. He packed away his kit and sat down to examine a mysterious little bundle that the store-keeper called a "Housewife."

He undid the tape and unrolled a blue denim holdall. There were two rows of pockets filled with packets of needles, skeins of thread, a pair of scissors, assorted buttons and a knob of wax.

The very look of this dreary equipment sent him into a fit of the blues. It reminded him forcibly that he was only a swab and not only would he have to sew on buttons and patch his clothes, but he would be expected to sit down, cut out and make new sets of clothing when his original ones wore out.

What a career to look forward to after his dreams of becoming a great engineer! Downhearted and fed up with everything, he bundled up his "housewife" and threw it into his bag, regretting that he had joined up and committed himself to a life of disappointment and frustration.

9

THE UNIFORM changed the boys from carefree individuals to bonded minions of the Navy. It was a change the instructors had waited for with some impatience as it placed a seal on their authority and gave them a greater sense of power and importance.

Training practice took a more realistic turn. Wheedling and mollycoddling went by the board. A cut of a rope's end, a nasty back-hander or a seaboot in the pants emphasised the orders of the day. Anyone who didn't "jump to it", lively as a flea, was blacklisted for showing a bad example to the others.

A number of the instructors were of the "Bluenose" type who had answered the call of the sea as boys and grown up in the Navy with a fanatical devotion to discipline and duty. These old shellbacks were obsessed by an idea that the boys would never become good sailors if their lives were not made as hard and unbearable as their own.

As a class, the "bluenoses" were respected and although their bull-dozing methods were condoned, they were not encouraged by the young officers of the new Navy. It was allowed however, that these morons were true patriots, who, despite their crude ideas, inspired their charges with loyalty, courage and pride in their profession and handed down a tradition of seamanship that enabled Yankee clippers to outsmart the ships of other nations and add laurels and prestige to the Stars and Stripes in every seaport in the world.

Winter months passed in dreary succession. Unless the weather was very bad, the boys "fell in" on the parade ground at sun-up and spent the morning at squad drill, twisting, turning and marching in every troop formation known to the Marines.

Cutlass and rifle drill in the afternoons was a welcome change from foot-slogging and the squad would have been reckoned a "happy" one but for a brutal "bluenose" instructor who seemed to have a grudge against all mankind.

"Now listenta me, yez sons o' bitches," he'd start off when the mood was on him and it wasn't what he followed up with, but the way he said it, that made every recruit feel as though he was a mangy mongrel slinking around the parade ground beaten and subjected. Nothing less than clockwork precision would do him. If anyone had the misfortune to bungle his exercise or drop his musket the whole squad would be ordered to "Shoulder arms" and double up and down the vast parade ground until collar bones ached and shoulders were skinned and bleeding.

When the weather was too bad for boat drill, or sailing in the cutter, the boys were sent over in a pinnace to the training ship 'Constellation' which was moored in the Bay. On this famous old man-of-war, they learned about hull construction and decks, the parts of the masts and yards, the stays and running rigging and elementary sailorising, such as hoisting or sending down a yard and setting or taking in a sail. After much pulley-hauling and dodging among the crowded ropes on the pin rail, they became slick enough to be able to grab the right sheet, halyard or buntline without having to think twice.

Climbing the rigging came natural to Nat and he showed Kerr the run of the ropes and taught him how to look after himself at a dizzy height.

"Like so," he explained, "always bes' hand fer yersel', udder fer de ship, hang on wid yer teeth in a blow. Der, dat's it, dat's it; so."

After a few weeks practice the lads could run up the rigging like monkeys, move out on the swaying foot-ropes to the ends of the yards, shake out or take in a reef, snug down, or tackle any of the routine tasks that comes the way of a deepwater sailorman any day at sea.

So skilful did they become that for a dare, most of the class would climb the rigging, up over the fore-top to the topmost shrouds, shin up the th'gallant mast to the royal yard, place one hand on the truck, grab the main-royal backstay with the other and swing down hand-over-hand over a hundred feet to the deck.

The frigate 'Constellation', 1263 tons, launched in 1797, was one of the last wooden warships to be commissioned by the United States Navy. The vessel was one hundred and sixty one feet long and measured forty feet across the beam.

Armament consisted of twenty-eight 24 pounders on the gun deck and twelve 12 pounders on the spar deck. Twenty guns could be brought to bear in a broadside. Her crew averaged three hundred and ten officers and men.

The 'Constellation' had a glorious record of service. Early in 1799 the ship was sent out to intercept French privateers which infested the Caribbean Sea off the coast of New Orleans and plundered American ships trading between the West Indies and the United States. One morning in February, the 'Constellation', patrolling off Nevis, one of the Leeward islands, was alerted by the lookout when the topmasts of a French privateer appeared on the horizon bearing down towards them under full sail.

Captain Truxton ordered all hands to action stations, guns were shotted and run out, powder brought up from the magazine, cutlasses and boarding pikes handed around among the crew. Then he lay-to to intercept the vessel, which on coming closer proved to be the 'Insurgente'.

When the privateer came within range, 'Truxie' sent a shot across his bows and ordered the Captain to strike his colours. The Frenchman's reply was unequivocal and in keeping with the spirit of the times. Luffing up into the wind like a small boat at a regatta and before the astonished Yanks had time to recover from their surprise, the 'Insurgente' let go a murderous broadside into the hull of the 'Constellation' at point blank range.

Then followed a short but savage fight as the ships closed in a death-struggle. The 'Constellation's' guns belched thunder, flame and metal and soon the decks of both vessels were slippery with blood. The privateer suffered enormous losses in men and her captain, realising that his attack had failed in its object, hauled off and tried to sail out of range.

'Truxie' assuming from the sporadic return fire and the cries and moans of the wounded that the battle was half-won, set all available canvas, overhauled and boarded the vessel with a detachment of Marines. With his decks like a shambles and over three quarters of his officers and men wounded, the captain struck his colours and was taken prisoner. The prize was then towed to St. Kitts and the victory celebrated in great style by the whole population of the island.

The tales that were told about that and other notable victories echoed down the years, long after the time when the 'Constellation' retired from active service and became a training ship.

"Here at number four gun on the port side, is where one of the gunners panicked and fled. This is the spot where ensign Clarke, in full pursuit, had both legs shot off by a ball from the 'Insurgente'. Gripping his pistol between his teeth and with the life-blood spurting from his shattered limbs, the officer dragged himself almost the full length of the ship and caught up with the deserter cowering against the forecastle bulkhead. Almost on his last breath, the avenging ensign aimed and shot the man dead; seconds before expiring himself."

The whole ship exuded an atmosphere of the past. The quarter deck, the gun deck, the sick bay and the magazine, the ward-room and the men's quarters, all had their anecdotes of heroism or shame.

It was said that the 'Constellation' was haunted by the ghost of an old shellback named Tom Bardon, who was caught stealing rum from the medicine chest when he was sent to clean the sick bay. In the Old Navy, rum was used to deaden the pain of operations before chloroform came into use. The miscreant was sentenced to twenty lashes of the cat-o-nine tails.

Officers were assembled on the poop, ratings on the after-deck. The prisoner, bared to the waist, was bound hand and foot to the mizzen mast. At a nod from the captain, the boatswain's mate ran his fingers through the thongs of the 'cat' and commenced the flogging. The unfortunate wretch cried out in pain until he lost consciousness at the fifteenth stroke. Although his bones showed white through the bloody mess of skin and muscle, the senseless lashing continued. Shortly after being taken down, the prisoner died in the sick bay. A haunting tale of brutal punishment in the Old Navy.

* * *

After passing his examination in seamanship Kerr was transferred to a class which marched over to the heavy gun battery at South point daily for instruction in gunnery. This was the first time he felt any real interest in his work since joining the Navy. To him it seemed like entering another world — the world of practical mechanics in which he hoped to make his career.

In a long line of sheds by the water's edge, guarded night and day by Marines was an assortment of naval guns. Some were mounted on modern carriages, others were housed in mock turrets. The secret weapon of the period was there, a huge 16-inch smooth bore gun, mounted on a carriage with hydraulic shock absorbing gear to take the devastating recoil when the gun was fired.

Looking back, it seems incredible that as long ago as 1894, the United States Navy had a gun capable of firing a shell weighing half a ton that would pierce an 11-inch thick nickel steel armour plate at a distance of over three miles. The gun rotated on a massive capstan and was so nicely balanced that it could be elevated or dipped and deflected to the right or left by manipulating a control wheel with one hand.

Kerr made full use of this opportunity for learning the mechanics of gunnery and soon became skilful enough to take to pieces and re-assemble some of the lighter pieces of ordnance such as Maxim, Parrott and Gatling guns and was first in his class to pass the test of dismantling and re-assembling the intricate mechanism of a Dahlgren gun, rated a headache even by veteran instructors. He also studied the new electric firing circuits which were at that time operated by primitive cells.

Time passed quickly at this absorbing occupation. The four, - six and eight-inch quick firing guns had a lethal significance about them, that stimulated their youthful enthusiastic crews. All took turn about at the different control stations on the gun. Opening the breech, loading the shell, closing the breechblock, elevating or depressing the gun to the orders of the gunnery officer, setting the range-finder, sighting and then firing . . .

It was all exciting and although there was no thunderous roar, no smoke nor recoil with dummy shells, the crew thrilled and cheered as though they had scored a hit on an enemy ship.

A week after passing his examination in gunnery, Kerr, Nat and a crowd of messmates stood at the notice-board near the flagstaff reading over and over again the names of those who had passed and were now rated ordinary seamen Second Class. Most of the lads were thrilled, but Kerr didn't get much of an uplift from the news that he was promoted to the rank of swab.

There was jubilation when it was announced that the men who had passed would be granted a week's leave before being drafted to form ship's crews. Those who had homes nearby received free transport and the remainder elected to stay at the Seamen's Mission in Newport. All the lads were now well developed, tough and in fine fettle as they roamed about Newport sporting Western ocean rolls as though they were old seadogs.

Saloons were out of bounds and Kerr resolved to abide by this order. He was walking around, seeing the sights with Nat, when they fell in with a few messmates who had broken out and started on a drinking spree. At that time a man or a boy in uniform could get liquor for next to nothing. Kerr was greatly humiliated when his friend Nat left him to go with the fellows on the booze, and for the first time began to realise the degrading effect drink had on young people. He never felt the same regard for Nat.

Shortly after the holiday break, Kerr and some others were detailed to make up the complement of a newly launched gunboat which was fitting out for sea at the Navy yard, League Island, Philadelphia.

The 'Concord' was a steel twin screw vessel of 1710 tons, 230 feet long and thirty six foot across the beam. Armament consisted of six 6-inch guns, two six-pounders, two one-pounders, two 37 mm. and two Gatling guns. Her complement was one hundred and fifty officers and men.

This gunboat,was one of the first batch of ships in the modern Navy, and Kerr felt a thrill of pride at being picked as a crew-member. From the first time he saw the modern warship on which he was to serve, he had a new outlook on life. Everything was so clean and bright — such a contrast to the dilapidated 'Constellation' and the wretched slums he had existed in since coming to New York.

Life on the 'Concord' was go-ahead all the time. The whole ship's company, from the Captain down to the galley-boy worked like a unit geared to a pitch of top-notch efficiency. It was here that Kerr became imbued with the spirit and drive of the men who were building America.

The teamwork at gun, fire and boat drill was wonderful and all routine tasks, down to the meanest chores in the ship were performed with care and devotion to duty. Kerr was amazed at the speed with which signalmen could connect up and send a string of flags fluttering to the truck with a code message and the slick way they could spell out, or talk at a distance in morse or semaphore.

All good-conduct men were allowed shore leave in the evenings and there was plenty of time at week-ends to explore the city of Philadelphia, which William Penn founded as a Quaker colony in 1682.

Most of the public buildings were open all day. Kerr went the rounds with sight-seers and visited the old house number 239 Arch Street, where Betsy Ross made the first American flag. Nor did he neglect his chance to 'see and learn' in the great industrial sections of the city.

Kerr and a party of friends toured the vast Baldwin Locomotive Works and Cramp's Shipyards on the bank of the Delaware River. It was an education in itself to see giant cranes in action, cantilever bridges heeling up for ships to pass, transporter bridges crossing the water at high level, great ships building on the slip-ways, the noise, bustle and activity of it all against the vast smoky background of the city held them spellbound.

10

ONE AFTERNOON the Blue Peter fluttered at the foremast head, the ship's company stood by at watch stations or lined the rails, the gangway was hauled up and with a throaty basso-profundo moan, that drowned the cheers of a host of well-wishers, the 'Concord' swung around and slipped slowly down river for trials at sea.

After casting off, the crew went on sea watches. At that time, watches changed every four hours, with two dog watches to break the sequence, the first from four to six p. m. the second from six to eight p. m., then the First Watch took over duty until midnight. Within a few hours the gunboat was in Delaware Bay, steaming out into the darkness of the north Atlantic on a course for Cape Hatteras.

Being a light craft designed for service in shallow waters, the vessel was slightly top heavy with barbettes and other superstructures. Once clear of the land the ship rolled to a frightening degree, dipping rail and rail into the sea and scooping up water, so that decks were awash for hours on end in the heavy ground swell.

Most of the watch were dry-land sailors, who quickly succumbed to seasickness and reached the lowest depths of demoralisation and despair as they held on for dear life or vomited and skidded out of control until suddenly brought-to, dazed and half-drowned in the scuppers.

Fortunately for Kerr, he soon got over his bout of sickness and was able to take his first trick at the wheel at noon on the following day. A quartermaster stood by, guided him occasionally and explained how to steady the ship on her course. The compass card swung so wildly with the motion of the ship, that Kerr took his advice to steer by keeping the ship's head steady on a distant lightship and to gauge the centre of the arc described by the compass card now and again to check his bearing.

On the third day out, Cape Hatteras was on the beam and the 'Concord' headed south at full speed into the Gulf Stream and fine weather. The best average test run was sixteen knots in one hour. When the speed trials were over, the crew settled down to the rather monotonous routine of life on board a warship in peacetime.

Navigation was carried out by the captain and his officers in the chart room and on the bridge. Engineers were seldom seen on deck except in fine weather. Stokers and coal passers (the black squad) worked out their watches down below, in the stokehold and bunkers. Marines acted as sentries and ship's policemen. Cooks and stewards fussed around with mess-kits and dishes and were usually to be seen darting around the galley at meal times.

Guns' crews cleaned and oiled the guns daily, until the barrels were burnished and glinted in the sun. Signallers seemed to be the most active members of the crew, running up colourful strings of flags at lightening speed, when a sailing ship wanted to have her position reported or to send a message to her owners. There was still a great number of windjammers on the trade routes when the 'Concord' made her first trip to sea.

Swabs outnumbered all other ranks and were mostly occupied with the Herculean task of keeping the ship clean and tidy: holystoning the decks at an early hour, cleaning the portholes, polishing brass-work and filling innumerable sanitary tanks in the officers and men's quarters.

Kerr felt out of place slopping about in his bare feet with a bucket and a handful of rags, washing what seemed to be acres of paintwork, especially if an engine room junior came up from below, gave him a disdainful glance and paced the deck a few feet away with an air of importance. The rumble and swish of the engines, the hiss of escaping steam, even the stench of overheated oil coming up through the fiddley, stirred him with a longing to be at the work he loved.

No one but the captain knew for certain where the ship was bound for, but on the fourth day out a rumour went around that the vessel was to put in to Savannah for bunkers and stores. No one was surprised when the 'Concord' altered course and steamed purposefully between the mud flats of the Savannah River estuary.

As they nosed through the sluggish yellow waters under a blazing tropical sky, lazy outward-bound steamers and sailing ships in tow passed by, laden to the marks with lumber, cotton and resin, filling the air with an aroma of turpentine and gum.

The flat country on both sides of the river spread out into a vast panorama of cream and green cotton fields, dotted with haciendas which dwindled in size as they faded into the shimmering haze.

It was afternoon before the 'Concord' tied up at the coaling wharf at Savannah, the first great naval stores port in the United States. All shore leave was stopped until the bunkers were filled

The following morning, officers and men turned out for first parade in their oldest or someone else's clothes, looking as odd as a crowd of school-boys on Guy Fawkes Day. Young sailors anticipated great fun; old shellbacks swore, spat tobacco juice and wished they were back in the good old days of sail. Officers volunteered to take part in the coaling and only a few ratings were exempt from the task.

Fires were damped down to avert an explosion; skylights, ventilators and living quarters were covered with tarpaulins to keep out the dust. Runways and guide ropes were set up on deck. At the dump on the wharf a few yards away, hundreds of baskets filled with coals were ready to carry away.

Officers directed operations. The men were divided into groups, each forming an endless chain. Once set in motion the chains of men worked with the precision of machinery. Shore gangs lifted the baskets; crew-men took them on their backs, along the runways and dumped them down the chutes and hatches.

Down the stokehold and in the 'tween decks coal passers and trimmers — the colour of piccanninies — worked like demons, shovelling and wheeling the coal in barrows to the furthest corners and stowing it so no space was wasted. In the sub-tropical heat, the 'tween deck was an inferno. The men perspired and gasped for air but everyone kept at it with a will and by evening, there were eight hundred tons of coal in the bunkers and the hatches were battened down.

It took two days to get the ship clean again. The coal dust had penetrated into everything. It was in eyes and ears, shore leave boots, kitbags, bedding, in the food and a week later when all was forgotten made unexpected appearances when something was moved, to remind them of the day they coaled ship.

One afternoon, a notice pinned to the board confirmed that the 'Concord' would remain in Savannah for three weeks and word got around that surveyors were coming down from New York to examine the ship, prior to it being accepted by the Navy Board.

Next morning at first parade, a call for volunteers to train for diving was made. Before the officer finished speaking, Kerr bounded from the ranks and after being questioned later, he and five other ratings were accepted for the course.

There were diving suits in the stores and after learning how to put one on and take it off again and to regulate the valves for breathing, the day came when everything was ready for the learners to take turns at going into the water.

Sitting in the shallow diving punt, with the water lapping coldly at the sides, Kerr felt anything but happy as he got into the clumsy rubber dress and strapped on the heavy brass boots. Two men were practising at the rotary pump and the air wheezed out from the tubes. The thought persisted in his mind, what would happen to him if the pump failed? When the corselet was settled comfortably on his shoulders, the heavy back-weights strapped on and a man standing by ready to bolt on the helmet, he was in a cold sweat and felt like backing out altogether.

Just then the Warrant Officer; a master diver gave him his briefing.

"Don't twist around suddenly and foul your air lines and don't lose your head whatever happens, just jerk the signal cord goodo an' we'll have you up in a jiff."

Before he had time to reply, the helmet was bolted on and Kerr began breathing air from the pump. Unceremoniously, he was half-lifted and guided to an iron ladder at the stern of the punt. He gripped the side rails and stood on a rung at water level. Looking up through the window of the helmet he could see a crowd of shipmates waving and shouting on the deck of the 'Concord'. There were no more steps down, then there was no support. He lowered himself gingerly by hand to the last rung and let go.

The water gurgled noisily and closed in over his helmet. His heart pounded as he went down, down, into the semi-darkness until he felt his boots sinking in the mud. Then he pulled the life-line twice to let them know he was on the bed of the river. He drew his long diver's knife, ready for anything.

When his eyes became accustomed to the gloom, he could see a yard or so ahead. Slowly and awkwardly he walked towards the hull until he could make out the edges of the plates and rivets. The water clouded up with silt at every step and he had to wait for it to settle and become clear again. His hands were numbed with the cold and the pressure of the watertight bands on his wrists.

As he moved towards the bow he discerned an object about five feet high a shade darker than its vague background. It appeared to sway and move towards him. Kerr got the jitters, lost his nerve and jerked the life-line frantically.

When he came to, he was seated in the punt. The diving helmet was off and he was recovering from shock. His nose was bleeding and he had a splitting headache. Decompression stages were unknown at the time Kerr learned diving.

"You've been down ten minutes an' done very well," said the instructor. "Take the rest of the day off an' you'll be O. K." Although his first attempt at diving ended ignominiously, Kerr had no intention of leaving the class. He took his turn regularly with the rest of them and, when the surveyors and underwriters arrived from New York was fit to go down with them and take part in the examination of the hull. During an inspection, he found that the spectre that had scared him was only a bale of cotton standing on end in the mud.

When he received a badge, rating him a diver, life on board the 'Concord' became more pleasant. Apart from a small increase in pay, Kerr gained status with his shipmates.

When the ship's company went on shore leave in Savannah town, most of them made whoopee as long as their money lasted. Some tried to prolong the carousal after their money was spent, got into trouble with the police and landed in jail.

Kerr had sense enough to know that if he wanted to get ahead he would have to cut short that sort of thing, even if it meant losing friends. He had learned that friendships formed in drinking saloons weren't much good to anyone.

As most of the naval personnel were of immigrant stock and knew little about their adopted country, all ratings were encouraged to learn about the history and development of the United States. Books and writing paper were issued free and self-education became the passport to positions of responsibility and trust.

Kerr got down to study and found the History of the United States most interesting. In 1865, at the end of the Civil War, there were seven hundred ships in the Navy. Sixty-four of these were ironclads, that is, wooden ships with armour plate just above the water-line.

As a first step towards creating order out of post-war chaos, the energies of the nation were concentrated on vast reconstruction schemes: rebuilding industries, laying down railroads, establishing banks, reviving trade and commerce and opening up new towns and cities from the forty-ninth parallel to the Gulf of Mexico.

The possibility of an invasion by sea was so remote that the Navy was neglected and the warships laid up and forgotten until they became dilapidated beyond repair. While the United States was preoccupied with its giant coast-to-coast peace programme, kingdoms and empires rose and declined in the rest of the world.

Britain gained control of the Suez Canal and opened up great trade routes to Australia and the Far East. Burma, Egypt and several countries in Africa were annexed by the British. The French and German empires flourished, grew in strength and strove to acquire more and more foreign possessions. The Japanese Empire expanded and became all-powerful in the Orient. Most of the nations were jealous of Britain's grasping imperialism and languished in fear of her Navy, which dominated the seven seas. A prolonged state of world tension started the first great armaments race.

In 1880 the United States Navy consisted of a heterogeneous collection of warships, half-steam and half-sail and its strength was reckoned about twelfth among the navies of the world. There were only forty-eight ships that could fire a gun, and of these, not more than thirty wooden corvettes and sloops-of-war were fit for service abroad. There were no steel works in the United States capable of making forgings for guns of more than 6-inch calibre, and armour plate, propeller shafting and machine guns, had to be imported from abroad.

Roused to action by this picture of national decadence and apathy, Commander J. King of the U. S. Navy, wrote a book called "Warships and Navies of the World." This book put the spotlight on national insecurity, stirred the public and the following year the Secretary for the Navy, Mr. Hunt, convened a Naval Board headed by Rear-Admiral John Rodgers, to consider the Navy's requirements.

The estimate was for thirty-eight new steel cruisers, twenty-four gunboats and torpedo boats. In 1883 Congress reduced the programme to four warships, 'Atlanta', 'Boston', 'Chicago', and the small despatch boat 'Dolphin'.

From that time on, however, the drive for a new and efficient Navy gained momentum. In 1885, Congress approved the construction of thirty-three men-of-war, including the 6,315 ton battleship 'Texas', which carried two 12-inch and six 6-inch guns and could steam 18 knots. Another notable ship of the period, was the 3,730 ton 'Charleston', the first U. S. warship to discard sail rig. In 1887 the great American steel corporations began to produce 16-inch naval guns and vast quantities of armour plate for building ships of war.

11

WHEN KERR was serving on board the 'Concord' at Savannah, the Navy was constantly in the news and he eagerly read the reports in the great New York newspapers, which were highly prized in provincial towns and were read and re-read as hot news, long after the date of publication. Growing up, mentally and physically, in pace with the times, he developed a broad outlook, or; as a Yank would say, "began to see things big."

Sometimes he would go on shore leave with a crowd of shipmates, but mostly he went alone. He never felt lonely.

Savannah, known in the eighties as the Forest City, on account of the great variety of trees that flourished there, seemed to him a wonderful and exotic place. The market, shaded from the sun by giant sweet-scented magnolias, was colourful and picturesque.

Big-breasted negresses with laughing faces and fat arms, clad in every conceivable combination of coloured cotton frocks and headgear, moved sedately between stalls of oranges, watermelons, peaches, mangoes and other tropical fruit, gesticulating and chattering, while impish piccanninies with gleaming eyes, flashing teeth and nothing else, played hide and seek between skirts and got in everyone's way.

Strolling on one of the fine tree-lined main streets one scintillating blue-black night, Kerr had his first experience of the race hatred he had heard so much about.

At first it seemed strange that he couldn't see a negro until he almost bumped into one. They emerged so suddenly from the darkness. A floating white collar and a pair of swinging cuffs materialised into a negro, padding by with a sort of animal grace and vanishing into the shadows again.

They came so close that it seemed like a challenge as to who had the right of way, or who should give way to the other. His sleeve accidentally touched a negro. There was a hiss and a voice said "White trash, white trash," as the negro spat vehemently and went his way.

Kerr walked back towards the 'Concord', shuddering as he remembered his shipmates' warning, that most negroes carry a cut-throat razor in the hip pocket and would be likely to use it with jungle ferocity on any sort of provocation.

With his Irish background Kerr had no animosity towards coloured folk, in fact after reading 'Uncle Tom's Cabin' he felt that the United States negroes had been inhumanly treated, but — as he often reflected later — those sentiments wouldn't have helped him one bit if he'd unwittingly "walked into it", on the sidewalks of Savannah.

Next morning there was great fun and excitement when the 'Concord' got unexpected orders to go to sea. Boatswain's pipes shrilled. Officers shouted commands. Last-minute stores were rushed aboard. Down below, shovels clanged on the stokehold floor, smoke billowed from the funnel, propellers throbbed, the ship pulsated with life. To the sound of a thousand cheers from a farewell gathering on the quay, the last hawser was cast off and the 'Concord' proceeded on her voyage.

The rumour went around that the vessel was under sealed orders. After clearing the river delta, the 'Concord' followed a southerly course through the Florida Strait, coasted along the chain of keys and atolls at the toe of the mainland and on into the Caribbean Sea. A course was then set for a position plotted on the chart as Lat. 25. N. Long. 85. W.; it formed part of a navigational exercise for the warship to meet a collier and to "coal ship" at this rendezvous.

On a sultry afternoon, two days after passing Key West, the naval collier 'Merrimac' was sighted and gave the recognition signal as arranged. The vessels slowed down and manouvered into position in a sea as calm as a mill-pond, closed in gently on fenders and made fast alongside.

The 'Merrimac' was scheduled to transfer a thousand tons of bunker coals to the warship. While the captain and officers were exchanging courtesies, the collier's derricks were topped up and loading, by means of winches and mechanical grabs began. This was an eye-opener for the men who were used to humping the coals in baskets on their backs, also it was a practical demonstration of the latest method of bunkering and the power and efficiency of steam.

By nightfall the coaling was finished and the ships parted company with much whistle-blowing, flag-wagging and cheering, as was the custom of the period. Shortly afterwards a notice was posted up to confirm that the 'Concord' was under orders to proceed to New Orleans to join a flotilla which included the protected cruiser 'Boston', battleship 'Texas', monitor 'Monaduck', gunboat 'Petrel' and other naval units in the port.

When the first hilarious round of visits to ships, get-together parties and fraternisation between officers and men of the flotilla fizzled out, interest naturally gravitated to the romantic city of New Orleans.

Built on a vast plain, one hundred miles up-country from the mouth of the mighty Mississippi River and spreading out mostly on its left bank, the city at first gave the impression of being a world unto itself, but on closer inspection seemed more like two worlds that were pre-ordained to merge, but never achieved their destiny.

The new American sector, which included public parks, administrative buildings, banks, churches, railway stations and a network of modern electric tramways, was separated from the old picturesque French city by Canal Street, a main throughfare. Negroes and creoles thronged the shops and sidewalks; white folk were in the minority.

In the eighties, New Orleans was a rapidly expanding cotton port with thriving industries, manufacturing cotton goods, machinery, boots, shoes, cigars, furniture and other products. How the city originated in such an unsuitable location has ever been a puzzle to modern engineers and town-planners.

Except for the modern city centre, which stands on a rise, the ground on which New Orleans and its environs are built, is well below the flood level of the Mississippi River. Fernando de Soto found this to his dismay, when the river overflowed its bank and flodded the countryside in 1542.

As the city grew and developed, massive embankments or 'levees' had to be built, to hold back the ever-threatening waters of the river. Despite this protection, the levees were often burst by unusually high storm-swept tides and the city and surrounding country became indundated by floods.

Another elemental problem that had to be overcome was that the low-lying marshlands on which the city spread out were so waterlogged, that it was impossible to bury the dead in graves. The burial problem was solved by placing the coffins on the ground and building the earth around and on top of them. The coffins were stacked over one another in layers so that in time a cemetery became a vast plateau twelve or fifteen feet higher than the original ground level.

It is unlikely that the city could ever have been built in the face of these great natural difficulties, had it not been for the never-ending supply of cheap labour, that flowed annually into the budding metropolis. This was partly due to the exigencies of the times.

Although slavery was abolished in the United States in 1862, the emancipation of coloured folk did not always work out in their best interests, and for that reason conditions of slavery persisted for generations later. Even though it was laid down by law that a plantation slave and his family had a shack to live in and were entitled to a dollar or so and an allowance of tobacco each week, most of the slaves were afraid to accept, knowing that if they took the money, they and their families would be evicted and hustled off the estate by the overseer.

Unlettered and unskilled and denied by the whites the right to learn a trade, the emancipated cotton-pickers were free only to roam the country and starve. Thousands made their way to New Orleans, hoping to find more lucrative employment, but when they got there, found there were ten men looking for every job.

Goaded by hunger and want, trump cards in the hands of employers and financiers, they were forced to sell their labour for a pittance and work out their lives building railroads, miles of levees and docks, compressing cotton for export and sweating on other vast enterprises controlled by multi-millionaire opportunists.

In the American business section, there seemed to be about three hundred coloured folk to every white person. Negroes and creoles crowded the sidewalks. Whites, smartly dressed, self-conscious and self-important, carried themselves with an air as though they were the ruling class by natural right.

Coloured men in cheap blue jeans and gaudy open-necked shirts, walked along with loose-limbed grace. Gaily clad, shapely creole women went leisurely about their shopping.

The carefree air usually associated with deep southern towns, however, was lacking, a state of tension born of racial animosity was evident. Coloured people, looked sullen and defiant, especially when whites drew near. When eyes met, glances of unconcealed hatred were exchanged as often as not. The atmosphere was smouldering and it seemed as though it only needed a spark to start an inferno of mob-violence.

Saloons were full of American sailors on shore leave at week-ends. All had been warned about getting into trouble and the danger of creating anything in the way of an incident, so they were mostly well-behaved.

Kerr liked wandering around the quaint old-world French part of the city. He was greatly intrigued with the rope-soled canvas shoes, offered for sale in the coloured folks' shop windows. They were for enormous feet, up to fifteen or sixteen inches long, evidently a racial characteristic handed down to negroes from generations of jungle-bred ancestors.

Everywhere there was evidence of a militant colour bar. Over the doors of saloons, hotels, theatres, cafes and shops were prominent notices: "Whites only", "Only Whites served", "Coloureds keep out", or "Coloured only"

It was not long before Kerr found out that coloured people were not allowed to hold positions of authority, except to supervise coloured labour gangs. All police officers and firemen, municipal officials, tram drivers, conductors and skilled tradesmen were white; only heavy laborious work was allotted to coloured folk. It was a generally accepted principle that a white could commit murder and get away with it if the victim was coloured. Coloured people were not called upon to serve on juries — in case they might be prejudiced against the whites.

Kerr considered that this was all very one-sided and discussed the matter at length with local aquaintences. From what he gathered, the "White front" seemed to be:

"Niggers must be kept down." "Keep them in their places." "Hammer them, whip them, string 'em up on a tree, it's the only way to keep 'em down." "It's all right the likes o' you talkin', but no white woman 'ud be safe if they weren't kep' down like that" they'd argue, or say: "We're in the minority. If ever they got the whip hand they'd jes' wipe us all out."

Kerr gave these angles much thought, but no matter what arguments were put forward in favour of the colour bar it seemed wrong, unfair and unconstitutional.

On a sweltering afternoon, about 90° in the shade and too warm for walking, Kerr boarded one of the new electric tramcars at the Union station terminous. It was a long, single-decked, roofed trolley car, with open frames for windows.

The slatted seats for forty passengers faced towards the driver's end and were divided into two sections by a gangway down the centre of the tram. A notice over the right-hand section read "Whites only" and over the left "Coloured". Each seat accommodated two passengers.

After paying his fare, Kerr got into the first seat behind the driver and had it all to himself. Passengers came in until the rest of the seats were filled. The car rumbled and clucked along Howard Avenue on iron-flanged wheels, screeched to the right at the intersection of Julia Street and headed north towards the old French quarter.

From the start of the journey, Kerr saw the extent to which racial discrimination could be applied even in comparatively small provocative ways. If white people crossed the road in front of the car the driver tapped the warning bell with his heel and slowed down for them to pass, but if coloured folk got in the way he didn't slow down at all, but stamped on the bell until it clanged like a fire alarm and the pedestrians had to run out of the way to save life and limb.

By the time the car stopped at Char Hospital, Kerr's Irish blood was up and his mind made up. In this matter of racial discrimination and segregation he decided to take sides with the coloured folk.

A hefty young negress nursing a baby boarded the car, walked up the centre aisle and turned to face the passengers, leaning against the partition to the driver's platform to steady herself and her charge. The only unoccupied seat in the car was in the white section, beside Kerr.

As the car moved off, Kerr stood up to make room and invited the negress to sit down in the corner.

"Here you are, ma'am, take a seat. Paid for one, haven't you?" he added encouragingly.

The woman rolled her eyes in amazement as only a negress can, then smiled back and replied in the melodious patois of her race:

"Lawsey, mister, I doant wanna make no trouble, but if you says so, sah, it mus' suah be all right."

She sat down in the corner. The car rattled on. A heavy silence fell on the passengers. For the moment everyone was stunned. Suddenly a row of grins broke out and relieved the tension on the "Coloured" side. Evidently everything was O.K. by them.

In the white section, however, the tension increased. There were agitated whispers and furtive movements. Kerr suddenly realised that he had created a precedent, which in the opinion of some of the colour-conscious fanatics aboard, would amount to sabotaging their sacred, natural-born rights. He looked ahead and made a quick decision. It was definitely time to leave the car — before he became a target for some crazy, homicidal white.

Cagey as a lion tamer making his exit, he rose from his seat and walked quickly down the car to the conductor's platform. The driver was braking for the turn to Beauregard Square and the old French quarter.

"Lookin' fer trouble, guy?" said the conductor smoothly. He was only a small man but looked as tough and wiry as an apache.

Kerr shifted his grip on the handrail of the swaying car and turned to face him, bis back to the sidewalk. He looked into eyes blazing with hate and contempt.

Before he could speak, the conductor had drawn back his fist and delivered a short and vicious solar-plexis punch into his unprotected midriff.

Kerr doubled up and landed on the sidewalk on his bottom, winded, sick and sorry for himself. A jabbering multi-coloured crowd closed in around him.

"Drunken sailor," someone said knowingly, then there was a titter and the curious ones, completely satisfied, lost interest and left him.

Kerr got on his feet and staggered back towards the riverside and his ship.

But news of the incident on the tramcar spread around the town and fanned the smouldering fires of race hatred into flames. After work, a white mob gathered in Rampart Street to plan reprisals on the negroes for flaunting the colour bar.

Worked into a frenzy by the inflammatory speeches of political hoodlums and hooded terrorists, the mob stormed up to Toulouse Street in the negro quarter. Here they dragged three young negresses from their homes, tore the clothes off them and hung them by the neck from street lamp brackets. Amid the cheers and jeers of the mob their living bodies were riddled with bullets.

Within an hour every Jim Crow in the district was out for vengeance, armed with his favourite close-quarters weapon, a cut-throat razor lashed to a broom-stick. The whole night long, negroes and whites surged up and down the street, in an orgy of bloodshed, killing, maiming and horribly mutilating one another until the road was littered with corpses and the hospital wards crowded out with casualties.

None of the American Navy men were involved in the riots, but the savage outbreak of civil strife was much debated and left an indelible impression on the minds of the officers and men of the flotilla.

It was generally felt that the perpetuation of the colour bar was unethical and unjust and completely out of line with America's much boosted principles of freedom, for all citizens of the United States.

12

AFTER A few weeks of manoeuvres in the Gulf of Mexico with the fleet, the 'Concord' was commissioned by the U. S. Navy and despatched to Key West for orders.

On arrival at this exotic pleasure island of the American 'eighties, the ship's company found the flags at half-mast and the town in mourning. A millionaire's son, leader of the exclusive Society set, had been lost in his yacht a few miles up-coast and the tragedy almost closed down the resort at the height of the season.

The 'Coraline', a modern steel sailing yacht with auxiliary engines, disappeared while cruising off the Ten Thousand Islands, an archipelago between Florida Keys and Marcus Point, with the owner Jacob Heinz, junior, and most of the chorus in the leg-show at the Trocadero on board.

The yacht was seen half a mile off a cay by sponge fishermen and it vanished before their eyes in a puff of smoke. The vessel may have struck a reef or a submerged wreck or the fuel tank could have exploded — only the placid, weed-strewn waters of Florida Bay could yield that secret.

When old Heinz, a millionaire railway car manufacturer, heard the news, he nearly lost his reason. He loved his only son and with the unrealistic attitude of one who has never known frustration or a real set-back in life, just couldn't believe he was dead.

The mystery must be cleared up; his son's body must be found. Apart from other things, there was a pile of insurance money involved and there was no proof of his son's death. Surely the State could help in a case like this. He contacted friends in the Navy Department and offered a huge reward for the recovery of the body, dead or alive.

The outcome of it all was that one morning Captain Rodgers of the 'Concord' received instructions to conduct a search for the yacht 'Coraline', believed to have foundered off the West Florida coast. The captain received the order with his tongue in his cheek and made no comment.

This was Kerr's first experience of the power of the Almighty Dollar and political pull in the United States, the Constitution of which insisted on equal rights and privileges for all citizens.

There was a lot of talk on board about the proposed search and soap-box orators, ever ready to sow the seed of dissension, took the opportunity to point out the iniquity of the authorities when it came to discriminating between the rights of the rich and the rights of the poor.

Most of the crew were aware that thousands of Americans died from exposure and want every year, and that their bodies were gathered up in refuse carts and dumped in communal graves and it seemed inequitable that when a millionaire's son died, the newspapers should make a fuss all over the country and the Navy could be ordered to search for his body.

Although there was a big reward waiting to be collected there was little enthusiasm for the search among the crew. The task of finding the wreck of a small yacht, carried here and there by the unpredictable currents of the Gulf Stream, seemed foredoomed to failure.

However, the 'Concord' steamed slowly along the coast and miles of dangerous coral and sand seabed were sectioned off and searched. Look outs stationed aloft scanned the shallows through marine glasses; a sounding party cast the hand lead from the chains and droned out the depths at regular intervals.

During the warm tropical nights the crew sampled one of the pleasures of a millionaire's cruise. The hammock room was deserted; most of the men slept on deck under a blue-black star spangled sky. Each indicator star of the fiery Southern Cross had its spotlight on the dark, oily surface of the sea.

The inky depths lit up with sudden phosphorescence as some deep-water prowler chased its prey, or a school of tuna blazed an incandescent trail towards the bows.

During a hot spell, discipline slackened and tempers became frayed. In the shallows, where the 'Coraline' had last been seen from a cay, the diving squad went into action to intensify the search. One morning as the 'Concord' lazed under a cloudless sky half a mile out from a string of palm fringed reefs, there was some excitement.

It was Kerr's turn to dive. He was lowered into three and a half fathoms of water, the greatest depth he could stand, and was trying his feet on the sandy bottom and adjusting his eyes to the pellucid green light when a sudden shadow came between him and the filtering sunlight. A monster fish, waving its pectoral fins, moved up close and stopped. A fathom away a great unblinking saucer of an eye stared at him coldly. Kerr at once identified the fish as a white shark, well known and feared in the Caribbean as a man-eater.

His only defence weapon, a diver's knife, was useless.

Then another shadow began to take shape in the dim distance. His brain began to function again as the streamlined body of a ten-foot swordfish lunged and buried a yard of its shaft in the shark's body. It must have penetrated a vital spot. The water churned as the shark, with one agonised heave, twisted over on its side, snapping the sword off short, then, as if paralysed, lay inert and made no effort to fight back.

But the swordfish, a demon fighter, continued its attack, reversing a few yards and charging again and again, with bull-like ferocity, savaging the stricken monster with its jagged, broken-off stump of a sword.

The diving crew in the punt had ring-side view of the fight and Kerr was hauled up. The tumult in the water was so great that it took them all their time to save the flat-bottomed boat from being swamped. Within an hour the carcases of the shark and swordfish floated peacefully side by side in a patch of golden Sargasso weed. A school of voracious barracuda had already begun their feast. The same afternoon the search for the yacht 'Coraline' and her missing crew was called off by the captain.

A fortnight later the 'Concord' docked and made fast at Baltimore, Maryland, for engine modification and the installation of torpedo tubes. The crew was paid off. Those who had no homes to go to were offered temporary accommodation at the Naval Academy, Annapolis, only a few miles away.

The Academy, established on the grounds of Fort Severn in 1845 was being enlarged under the direction of Secretary of the Navy, William C. Whitney, who had earmarked it as a headquarters for training the vast number of technicians required for the great switch-over from sail to steam. Construction of ancillary buildings was going on day and night when Kerr and his shipmates arrived.

The Administrator, Commander Stephen Lowe, a rising genius, had just inaugurated a new "Think for Yourself" era among naval personnel and as he strove to encourage initiative and self-confidence he soon had a large and enthusiastic following.

He aimed to turn out men fit to keep pace with the times, to master complex machinery, gears and guns, in modern warships and to change the old-type rule-of-thumb officers into efficient directors, capable of solving complicated tactical and strategic problems of naval practice and warfare.

To attain this ideal, the Academy was laid out on the lines of a technical school and officers and men had every facility to become expert at any trade. There were shops for armourers, blacksmiths, plumbers, electricians and schools for torpedo men, engine room mechanics and ship fitters.

The college was in fact a marine university, geared to the tempo of the times and the vast industrial development of the United States. Kerr was energised by the bustle of the place. Instinctively he felt he had at last come in contact with the world of his ambition and in which he hoped to achieve success. Inspired by his yearning for knowledge he obtained permission to visit the machine shops and took full advantage of this privilege. The more he learned about mechanics and the abounding opportunities that existed for progress and promotion in this field, the more he felt downhearted about his rating as a swab.

One evening he walked across the empty parade ground to the signal mast, the assembly point for general parades. He stopped to read the schedules on the notice board, casually curious. There were notices about duty watches, parades, fire crews, meetings in the Lecture Hall and so on. Then straining his eyes in the semi-darkness, he read a notice in faded blue typescript. It ran:

"Any seaman or stokehold rating between eighteen and twenty-five who wishes to train as engine room operative, report to the Engineer Lieutenant at the Section office 'G' block on Friday morning." The notice was over two months old and had probably lapsed. Just his luck!

He couldn't sleep that night. Next morning, feeling scared and almost sick with apprehension, he knocked at the Section office door and asked to see the officer in charge. Lieutenant Frank Rogers was there. Kerr was shown in and explained the reason for his visit.

"Wait a minute," the officer replied, reaching for a file. "Oh yeah, I remember. Bit late, aren't you? Don't think we had any applicants for this, or if we had, they weren't suitable."

"Well, Sir, I'd like to get a chance at mechanical work," Kerr said.

"Yeah, a lot of fellas would, but have you served any time to a trade, or anything like that?" replied the officer.

"I've been recommended for dismantling and assembling light guns during training and I've done some diving Sir," replied Kerr hopefully.

The Lieutenant appeared to be more interested.

"Here Mr. Haslett," he said abruptly, addressing a Warrant officer. "This young man seems to have a flair for mechanics. I think he's worth a try. Take his name and get all particulars. If he's recommended by the Captain of his ship, make arrangements for his transfer. We need young men like him in the Engineering Department."

Kerr left the office and skipped across the parade ground. It seemed as though the whole world had suddenly come within his grasp and that his great ambition was about to be fulfilled. That evening he wrote home to make his mother happy with the news.

A fortnight later Kerr was transferred to the marine engineering section in 'G' block. He was rigged out with a smart fore and aft blue uniform, a peaked hat and a golden propeller arm badge, to indicate that he was an engine room artificer. His feeling of inferiority vanished with his bell-bottom trousers and little dicky hat.

Kerr was immensely proud of his advancement to a higher rating and resolved to live up to the standard of conduct that obtained in this specialised branch of the Service.

His mechanical training began in the torpedo school. In the 'eighties the old British eleven foot by fourteen inch diameter Whitehead torpedo was used in the United States Navy. Manufactured under licence by the Hotchkiss Ordnance Co., New York, the torpedo was driven by an oil-fired steam engine and carried an explosive charge of eighty pounds of T. N. T. It had an effective range of four hundred yards.

The Navy had a particular interest in these weapons at the time, as Lieut. J. Howell, U. S. N., had perfected a gyroscopic steering device that raised them from blunderbuss efficiency to modern devastating war machines, that could be aimed with deadly accuracy.

The multifarious parts of the weapon were manufactured in New York and assembled at Washington Navy Yard for security reasons. Components were also assembled at the Naval Academy, Annapolis, as part of the scheme for training torpedo crews. Kerr was issued with a steel locker and a set of mechanic's tools which he prized very much. He soon adapted himself to his work and became dedicated to it.

During his term at the Naval Academy, the accepted philosophy of the trainees was, "that most normal men have a weakness of one kind or another and the guy who had no faults or vices — or wouldn't admit to having any — well there must be something wrong with him." Those who didn't agree with this dictum were put in place by the majority.

Kerr attended the torpedo school daily and having a natural bent for engineering had no difficulty with his work. He soon acquired a working knowledge of the main parts of a torpedo and the day came when he could assemble an engine with its driving gears and propeller shaft, mount the installation on a test bench and get it in running order.

Section by section he mastered the ancillary installations, becoming expert at assembling and installing the delicate pendulum balancing equipment, that regulated the rise and fall of the torpedo as it cut through the waves and the gyroscopic steering device that held the weapon on a true course after it was fired.

Within a couple of months he could fabricate a "tin fish" by himself and tune up the control units until they worked as smoothly as a clock. After proving his ability he was entrusted with the dangerous, but coveted task of loading the explosive chamber with a charge of T. N. T. and fitting a fulmanite of mercury detonator in the warhead to make the deadly weapon ready for service.

When he finished his term of apprenticeship, Kerr passed his technical examination and was sent to join the crew of the newly-launched gunboat 'Petrel', at Baltimore as torpedo crewman.

The warship was a steel, single-screw vessel, 176 ft. long, 31ft. beam and 892 tons displacement. It was manned by one hundred officers and men. The 'Petrel's armament consisted of four 6-inch guns, two 3-pounders, one 1-pounder, two 37-mm. and two Gatling guns. It was one of the first series of warships to be equipped with torpedo tubes on deck.

In comparison to the struggle for existence that was the lot of wage-earners in every town and village ashore, life on board the gunboat was secure and free from care, although it was conceded that the vessel made dirty weather of a breeze and would almost stand on end in a blow.

Accommodation was spartan but clean, and the food was reasonably good although most of the crew said it was ruined by the cooks in the galley. Routine was normal for a warship in peace-time, a continuous round of parades, fire and boat drills, endless target practice with light and heavy guns and in Kerr's section, firing and retrieving torpedoes. Rigorous discipline was maintained while the ship was at sea.

Most of the activities on board were directed towards training personnel as part of the mammoth programme of expansion recommended by Secretary of the Navy, Benjamin J. Tracy, and sponsored by dynamic Senator Hall. For almost two years the 'Petrel' steamed up and down the Atlantic seaboard, from Portland, Maine, to Wilmington, Delaware, braving the harsh extremes of temperature, fog, blizzards and mountainous seas for which that stretch of coast is notorious.

Kerr suffered a lot at first from exposure to the elements but in time became seasoned and inured, to working on deck when the ship was like a half-tide rock. For attention to duty and efficiency at his work he was promoted to Leading Torpedo-Man and lived on the best of terms with the officers and men in his section.

After a marine survey and overhaul at Newport News, the 'Petrel' was sent on a good-will tour of the West Indian islands by the U. S. Government. In the 'eighties, the islands were divided into two main groups, the Greater and Lesser Antilles. Many of the islands in the vast archipelago were merely uncharted cays and atolls.

Most of the forty-odd large islands were inhabited by European administrators and settlers, negroes and creoles with a sprinkling of Indians and the almost extinct Caribs. Inter-racial marriage was uninhibited and there was no colour bar, although a man's chances in life and social status were usually assessed by the amount of white pigment in his skin.

The languorous, tropical climate made the people indolent and lazy but there was no necessity for anyone to work, as a bountiful Nature had provided everything for the simple life, free. As world events proved, the boundless natural resources of tobacco, sugar, fruit, coffee and spices could not forever escape the clutches of grasping foreign powers and one by one, approximately half of the islands were annexed or by some insidious means brought under the hegemony of British, French, Spanish and Dutch colonial expansionists and the remainder were acquired or bought over from corrupt governments by crafty American opportunists.

In the course of time the West Indian islanders awoke to the fact that their lands were owned and controlled by international commercial filibusters and that willy-nilly, they were shackled to a world economy that took everything they could produce, and gave nothing in return except hard work and the introduction of a spurious civilisation. In practice, the advent of the white man transformed their Utopian heritage into hotbeds of industrial slavery and discontent.

Too indolent and segregated to organise strong political fronts, the once-happy carefree islanders lived in a state of smouldering animosity towards their masters occasionally manifesting their feelings by sporadic revolutions and changes of governments that never seemed to have any security of tenure.

The 'Petrel' visited the republics of Haiti and Santo Domingo and then proceeded to Martinique, a French possession in the Leeward Islands which was later the scene of a disastrous volcanic eruption in 1902. At all these places the warship was received with ceremony and respect and the officers and crew were fêted ashore.

From roughing it during his early days in New York and travelling later at Uncle Sam's expense, Kerr, at the age of twenty, had acquired more knowledge than most young men in their formative years. Rid of his inferiority bogey, he made friends among his own class, and like the rest of them smoked a cigarette and drank beer when he was ashore.

In those days there was a naval tradition that shipmates spent their money in their own company ashore and anyone who didn't fall in line was soon made to feel that life wasn't worth living. The tradition was ingrained and few had the strength of mind to ignore it.

When the 'Petrel' lay at anchor in the picturesque harbour of Fort de France, overshadowed by the massive bulk of Mont Pelée, Kerr's fortune took an unexpected turn. Many of the officers were visiting friends on shore.

The crew was engaged on usual harbour duties, swabbing decks, painting the ship's side and cleaning brass. Signallers exchanged flag-talk with their contemporaries in the coastguard station ashore. Boat crews were out in the harbour practising for a regatta which was to be held in their honour.

Down in the almost deserted engine room, in sleeveless singlet and dungaree trousers, and with the corner of a sweat-rag held between his teeth, Jerry Wainright, the Fourth Engineer, was cutting out a section of condenser pipe that was choked with brine deposit. Although fires were banked down, the engine room was sizzling hot.

It was awkward overhead work and Jerry had to stand on an iron handrail, between steam pipes, to get at it. He used a file and hacksaw alternately, sweltering from the exertion and lack of air.

As he cut through the heavy-gauge copper pipe a shower of filings sprayed the back of his neck and lodged between his shoulder blades. Tormented with perspiration and irritation by the filings he rubbed the itchy part frequently with the sweat-rag, as far as he could reach.

He had the section of pipe cut just at knocking-off time and, too exhausted to go to the messroom for a meal, or even to wash himself, went to his bunk and turned in.

Next morning he couldn't get up and the chief engineer, seeing that Jerry's back was swollen and his skin was a dirty green colour, told him to lay off duty. The ship's doctor came to see him in the afternoon and diagnosed blood poisoning from copper sulphate entering his bloodstream through his skin.

There were no powerful antibiotics in those days to check the spread of an infection. The tropical heat aggravated Jerry's condition and he died in agony a few days later.

The following Sunday, instead of taking part in a regatta, the ship's company attended the funeral of young Jerry Wainright, Fourth Engineer of the U.S. Gunboat 'Petrel'.

Kerr wasn't surprised when the Chief sent for him and asked him if he would like to switch over to engine room duties.

The officer pointed out that there was great scope for advancement and that Kerr would get help and instruction from the watch engineers until he became proficient. He also mentioned that Kerr would be appointed Fourth Engineer and receive a rise in pay from the day he accepted the post.

Kerr agreed without giving the matter a second thought and started work in the engine room on the same day. He had already gained a fair knowledge of the working of steam engines from shoptalk with the engineers off-duty.

At first he found it a great change from life in the fresh air on deck and had to try hard to get used to it. In due course the 'Petrel' took on bunker coal, fresh water and stores and sailed for the island of Curacao, off the north coast of Venezuela, South America.

The Second Engineer, Warrant Officer Jeb Wilson, was a fine middle-aged man of the school-master type and Kerr had difficulty in following his instructions.

"Keep an eye on that bridge telegraph," said the 'Second' solemnly. "Never know the minit it'll ring. If it rings, jump to it an' keep yer hand on th'throttle. Safety of the ship an' everyone on it may depend on how quickly you act."

"Sure thing, Sir," replied Kerr, now fully appreciating his new responsibility.

The engine room was blistering hot as the 'Petrel' steamed through the Caribbean Sea en route for Curacao, the sea-water temperature often registering over 90° Fahrenheit when taken at midnight. With all the discomforts, Kerr felt on top of the world as he walked around the footplates checking the bearings, or climbed the steel ladder and along the catwalk between flashing eccentrics, to oil the whirling governor, which cut down engine speed when the ship slid into the valley of the sea and the propeller, half out of the water, started to race.

The stench of burning oil nauseated him and he lost his appetite for food, but that seemed a minor matter compared with the thrill of responsibility he felt, when the 'Second' went off for an occasional spell and left him in charge of the engine room.

"These ol' triple expansion engines are so reliable that they practically run themselves," said the 'Second' one morning, "but look out for a hot bearing, especially in the tropics." He showed Kerr how to test the heat of an axle with his hand.

"If a bearing gets overheated and seizes up, it means stoppin' the ship and explainin' to the captain, so don't let that happen, whatever else," he admonished.

Trying to keep a conversation going above the rumble-a-rumble of the engines and the swish of escaping steam, was like a shouting match at first, but after a while Kerr acquired the proper pitch of speech and the knack of lip-reading, so was able one by one to learn the secrets of how the dynamo generated electricity for lighting the ship and driving the pumps.

The 'Petrel' arrived at the island of Curacaõ on schedule and dropped anchor in St. Anne's Bay to a thunderous salute of guns from the Dutch fort on the mainland.

When the excitement of arrival in port and the novelty of trying to converse with foreigners wore off and ceremonial parades, receptions and visits began to pall, the ship's company, knowing from experience that when the money is spent, most good things come to an end, were glad enough to settle down to normal working routine again.

Except for the intermittent click of a hammer and chisel in a dark corner, the engine room was unnaturally quiet. Sunbeams, filtering through the open skylight, outlined the massive engine bed and framing and touched up the flywheel, crankshaft, pistons and conn-rods with an oily burnished glint. The huge assembly of wheels, pipes, cranks and shafting looked like the innards of a sleeping Bessemer giant.

As they stood under a ventilator shaft to get a breath of fresh air in the stifling noonday heat, Kerr told the 'Second' of his ambition to become an electrical engineer. The 'Second' became thoughtful, then replied:

"Well, Kerr, a ship like this is really a power station on a small scale. The coal comes aboard and is burned to raise steam in the boilers. The power of steam drives the engine and the dynamo is run off the engine by means of a belt. The dynamo generates electricity for the lighting installation and pumps. If you understand the working of this machinery thoroughly, there's no reason why you shouldn't rise to a high-up job in a factory or power station ashore, when you've had enough experience. I'll help you all I can." And he did.

After a pleasant stay in Curacaõ the official goodwill tour of the islands ended and orders were received for the 'Petrel' to proceed to Cartagena in Colombia on the north coast of South America for bunker coals, and to continue the voyage on to Nicaragua in Central America.

Apart from a sizzling electric storm off Santa Marta, when ball lightening played hide and seek between the foremast truck and the yardarm, the voyage was uneventful and the gunboat arrived at Greytown, the capital city of Nicaragua, on schedule.

Instructions awaited Captain Rodgers there to carry out a survey of the San Juan River for a proposed ship canal, to link the Caribbean Sea with the Pacific Ocean. Interest in this route had been revived by a commission set up by the U. S. Government, as the construction of the Panama Canal by Ferdinand de Lesseps, the French engineer had failed and was abandoned for lack of funds.

When everything was organised the survey of the river-bed started. Sailors were out on the river all day, working out pre-arranged areas in small boats, taking and recording soundings with the hand lead.

Although the climate was reckoned mild and temperate there were torrential showers of rain and the medical staff soon had a sick bay full of men suffering from chills, colic and other complaints.

All hands were tormented by mosquitos day and night and during a spell of particularly warm weather seventeen members of the crew were stricken with yellow fever. They were at once transferred to hospital in Greytown.

It was not uncommen at the time for whole ship's crews to be landed at this port with "Yellow Jack" and the Franciscan Sisters in the hospital were experts at treating the disease. Everything was done to aleviate the sufferings of the men but within a fortnight five of the seamen who were exposed to mosquito attacks in the open boats died, and were buried ashore.

Seven deck-hands, two coal passers, two officer's servants and Kerr, battled it out and survived. But the ravages of fever took a lot out of the men and they returned to the ship, shadows of their former selves, stamped with the saffron tint from which the fever got its name.

After a consultation with the medical authorities in Greytown, the ship's doctor advised Captain Rodgers to postpone the river survey to a more favourable season as the risk of infection and mortality rate would reach its peak during the current month.

Following cable representations to Naval headquarters, the captain received instructions to abandon the survey for the time being and return with his ship to base in New York. The voyage of sad and glad memories ended, when the 'Petrel' docked in the Navy-yard Basin at Brooklyn and the crew was paid off in accordance with the procedure of the times.

13

AFTER THE 'Petrel's crew was paid off, Kerr was granted a fortnight's leave, with instructions to return to duty as a member of the dry-dock maintenance staff at the Navy Yard, Brooklyn, N. Y. It was in the fall of 1890. The thought that he had already served three and a half years of his seven year term with the Navy filled him with amazement at the flight of time.

When searching around for a place to live in, he came to a big brown-stone apartment house in Pineapple Street, about a mile from his place of employment and a few blocks away from the East River water-front. Pat O'Driscoll, the landlord, a retired street sweeper, opened the door to his knock and eyed him discreetly. When he heard that Kerr was a shore-based naval rating with steady pay and an Irishman too, he relaxed, smiled and answered:

"Guess you're O. K., lad, but you'll havta share a room with three other guys. If that'll suit, come right in."

Kerr followed him and looked over the room. It was clean and tidy and there was no smell of bugs. He knew instinctively that this was better than the average rooming house.

"It'll cost you three an'a half bucks a week an' you'll be treated as one of the family, can't do more'n that," added Pat.

Kerr booked his digs and brought his gear down the same afternoon. At tea he met Mrs. O'Driscoll, a bustling little Cork women, her son Mick, who was going to join the Police force and her pretty daughters Maureen and Una. It was a homely, cheerful family and Kerr liked the place from the start. Next to him at table was a room-mate, Jake Holson, who worked in a local brewery. The food was plain but plentiful and good. Pat O'Driscoll told them about happenings during his street-cleaning days and nights in New York, always starting off with, "One day when I was working for the City —"

Kerr went to bed that night relieved to find that the place was free from vermin and that Jake and his two other room-mates were clean, decent workmen who were easy to get on with.

When his leave expired and he reported back to the Navy Yard for duty, Kerr found that he had the option of continuing on as a steam engineer or starting in the electrical section as he had about the same amount of experience in both occupations. The officials told him there was a great shortage of skilled men in the electrical profession and that there was a night school at the Yard, where ratings could study when off duty. Kerr decided to choose an electrical career.

He already had enough practical experience to qualify as an electrician (second class) and after becoming officially rated, was sent to work in the dynamo and transformer shops. Eager to make the grade, he attended night-school regularly and worked hard. Within three months he was promoted to electrician first class and sent to work on the ten-thousand-ton battleships

'Oregon' and 'Massachusetts', which were in dock to be fitted out with complete electrical installations.

Kerr settled down to work, satisfied with his homely 'digs' in Pineapple Street. Pat O'Driscoll was a broad-minded man who hadn't become case-hardened or embittered in the struggle for existence like most New Yorkers. He was a natural-born lay apostle, who influenced those around him for their good and never failed to remind them of their duty towards their less fortunate fellow men. Pat took a fatherly interest in Kerr and got him to write home regularly. The two daughters, Maureen and Una, already matured enough to be on the look-out for husbands, made a great fuss of Kerr. In all, Number Seventeen Pineapple Street was a pleasant place to live in.

Kerr and his room-mates thought themselves fortunate indeed to have got into such a decent house, for only a block or two separated them from a vast area of malodorous, vermin-ridden tenements that spread out into the Flatbush district, a notorious slum section, where families lived in an appalling state of degradation and filth.

The electrification of the battleship 'Oregon' was a job on which Kerr gained considerable technical experience, for he was in on the start. He learned how to lay cables for lighting and power, how to work out the sizes of wires to carry the different electrical loads and the technique of distributing the current to the various parts of the ship. He also learned how to plan electrical circuits and to install fuse boards, to prevent overloading and potential fire hazards.

There was such a shortage of skilled men in those pioneering days of experimental electrical conductors, insulators and fuses that high-ranking engineers and inventors did not think it beneath their dignity to put on a suit of overalls and work with their men, to ensure that everything was technically sound.

Kerr's senior officer on the 'Oregon' was Lieutenant Mark Westinghouse. Mark was noted for his technical ability and got the best from his men by encouraging instead of abusing them. He gave unstinted help to those who wished to get ahead in their profession. Kerr was obviously so ambitious and anxious to learn that the Lieutenant coached him at his work and encouraged him to strive for an executive position in the Navy.

In the U. S. N. of the 'eighties, naval discipline wasn't rigidly enforced, nor was the difference in rank a bar to friendship, when officers and men were ashore in civilian clothes. Mark and Kerr shared an enthusiastic interest in some of the colossal civil engineering projects that were going on in New York and became firm friends. Often they crossed over to Manhattan and travelled up-town, to study the layout of the city and see things for themselves and to try to solve some acute civic problem; such as the supply of drinking water for the millions, which was always a headache for the municipal authorities of the day.

They were attracted and fascinated by Broadway, one of the world's show places, which stretched the full length of Manhattan Island, dividing it roughly in half. In 1890, it was reckoned the ugliest and most overcrowded highway in America.

Built by uninhibited get-rich-quick opportunists to no pre-conceived plan, except that the buildings lined up with the sidewalks, the thoroughfare was a long series of inelegant contrasts. With the exception of the Astor mansions on Fifth Avenue and a few banks and churches, there was no building on it with any claim to architectural style or beauty.

A centre-city impression was one of massive department stores or hotels that occupied a whole block and in between them, stretches of jerry-built shops, poolrooms, beer saloons, theatres, funfairs and other garish establishments, which provided cheap entertainment for the millions who thronged its sidewalks day and night.

Although it was over ten miles long and proportionately wide, Broadway was never able to cope with the tide of traffic that flowed from its precincts. This condition worsened as the city expanded and the volume of traffic increased.

In 1890 one and a quarter million New Yorkers lived in Manhattan as closely packed as sardines in a can. In addition to these, a floating population of almost two millions, came in to work from outlying districts and satellite towns across the East and Hudson Rivers, each day and returned home in the evening.

During daytime the thoroughfare was crammed with slow-moving horse-drawn omnibuses, carrier's waggons, drays, broughams, landaus and cabs and the sidewalks were thronged with a steady flow of pedestrians. When thousands of factories, construction works, shipyards, commercial firms and offices closed down for the evening, every street in Manhattan was invaded by crowds that appeared to be afflicted with rush-mania.

The pace of vehicular traffic slowed down to a crawl, as the struggling multitudes packed into elevated trains and horse-omnibuses and piled on to drays or any kind of vehicle that afforded a lift in the right direction.

As the human flood surged through the streets in a free-for-all scramble for transport, waves of jostling humanity inundated the railway stations and milled about the platforms, waiting for trains to rush them off to their destinations.

The big stampede overran riverside landing stages and swarmed aboard ferries for Hoboken, Jersey City, Staten Island and Brooklyn. For two hours millions pushed, fought and elbowed their way towards anything that would carry them home, then the exodus eased off and the overworked transport systems slowed down almost to a stop. At dawn the gigantic network of early morning trains, passenger ferries and transport systems slipped into gear again and the mad rush to work started afresh.

People lost an enormous amount of time travelling to and from work. Like many others interested in public affairs, Mark and Kerr spent hours at the Grand Central railway station and other vantage points, watching the frustrated crowds trying to get home and wondering why nothing was being done by the civic authorities to improve the ways and means of doing so.

One thing they were sure of and that was that most of the peak hour traffic chaos was caused by slow-moving horse drawn omnibuses that pulled over to the sidewalks to collect and disembark passengers, and that the situation would remain so until they were taken off the streets. Public opinion was aroused and the daily newspapers carried glaring headlines about traffic jams, and dismal prophecies about what would happen in New York if the passenger transport tangle was not speedily unravelled.

The problem was so big that it was beyond the scope of any individual to tackle it. The provision of more ferry boats and the construction of bridges across the East and Hudson Rivers, eased the situation for a while, but the ever-increasing volume of traffic baffled all attempts for a satisfactory and permanent solution.

At last, some of the go-ahead civil engineering corporations got together and after a great expenditure of labour and expense, formulated plans for the introduction of a system of subway trains, similar to the Metropolitan Underground railway that was working successfully in London and submitted them to the City authorities for approval.

'Boss' Croker, the New York City Chamberlain and most of the Tammany Hall leaders were satisfied with the plans, but when they were put forward for official sanction, a group of millionaries headed by the Astors and Vanderbilts, who between them owned most of Manhattan Island, protested that the subways would undermine their palatial homes and vast business premises, so in face of opposition by these formidable representatives of vested interests, the plans were for the time being, rejected.

After this setback, another group of transport technicians obtained permission to try out a passenger transport system, known as 'Cable Cars' similar to one which was used with some measure of success in Highgate, London, in 1884.

The cable car system consisted of a wire rope or cable housed in an underground conduit, between a pair of rails set in the surface of the street. The cable was kept moving continuously, like an endless belt, by a powerful stationary steam engine at one end of the route.

As the cars could travel only one way on each line of rails, two parallel lines were necessary on each route. A continuous slot, protected on each side by a steel strip let into the roadway, was arranged on top of the conduit and each car was provided with a steel plate known as a plough, which passed through the slot and gripped the cable by means of a device, which could be opened or closed by the driver. Pulleys were arranged along the conduit to support the cable.

The first line of cable cars in New York city was installed in Tenth Avenue, which started at West Fourteenth Street and ran parallel with the Hudson River and on into Amsterdam Avenue. Kerr and Mark Westinghouse were among the crowd of enthusiastic spectators when the line was opened for service in 1890. After a few months running, it was found that a vast amount of power was wasted in keeping the cable moving. The supporting pulleys were very noisy and cost a lot to grease and maintain in order. Accidents were frequent at certain points of the line.

But the combined services of Elevated Railways and Cable Cars did not speed up the traffic sufficiently, to abate the ever-increasing time loss in cross-town journeys, and experiments made by transport experts in other big cities were watched with interest.

In 1885 an engineer named Leo Daft opened a three-mile electrical tramway in Baltimore, Maryland. He used an electric locomotive and a live centre rail to supply the power but the line did not succeed and was replaced by horse-drawn cars. Daft then tried out an electric overhead wire system successfully. In the late 1880's a number of small transport systems, electrified with the Daft system, were established in American towns.

Another transport engineer of the period, Frank J. Sprague, became interested in the electrification of the London Underground railways and returned to the United States in 1885 with the idea of introducing a similar system in New York. He contacted Thomas A. Edison, who was keenly interested in mechanical transport. After forming a company, they worked together on plans for an electric railway. As the project developed they argued and fell out over some technical problems and finally Sprague broke away and started his own company. In 1886 he received a contract to electrify the Richmond, Virginia system, and it was successfully run in 1888.

The general electrification of public transport systems was now under way in many American cities and the daily newspapers clamoured for a similar system to be established in New York.

Backed by the mounting pressure of public opinion, the directors of a large engineering concern had plans drawn up for the construction and development of an electric tramway system, to serve New York and once again proposals were submitted to the city authorities for approval.

The plans were examined by the City engineers and placed before a special Meeting of the Public Transport Committee in Tammany Hall. When the scheme was put up for acceptance, the proposed method of distributing power by means of overhead trolley wires was criticised and after a prolonged discussion the committee decided that owing to the danger of high-voltage wires breaking and falling on the crowded streets, it would ban the use of overhead power lines in Manhattan Island for all time. This ban had a paralysing effect on the efforts and enterprise of rapid-transit experts and for many years later the traffic problem in New York remained unsolved.

In the eighteen-nineties New York was an unsettling place for any young man to attempt to make his career. The city abounded with attractions, excitements and temptations that were well-nigh irresistible, to those who had money to spend and lacked the steadying influence of a strong will or home and family ties.

Daily newspapers were full of stories about men who started with nothing and by the development of personal characteristics, or making full use of opportunities, amassed fortunes and became millionaires. These stories, written by inspired and gifted journalists, made the accumulation of wealth seem so easy, that reading between the lines, one gathered that anyone with the same initiative, enterprise and tenacity of purpose could do likewise. This was one of the colourful mirages that sustained, stimulated and bewitched New York's struggling millions.

At twenty years of age, Kerr was satisfied that he was well on the road to becoming an electrical engineer, but when exchanging notes with fellows of the same profession in New York, he began to wonder if he was making the best of his opportunities.

One evening, in an up-town saloon, Mark and Kerr got into conversation with an electrical contractor who was engaged on piece-work lighting installations in apartment houses, and was taking full advantage of the boom when householders were changing over from oil and gas, to electric lighting.

"Youse guys on dem Goddam battleships is workin' for buttons," he said, downing his beer. "Yez 'ud be worth a hundred bucks a week each to me," he continued, "don't be suckers, skip th' Navy an' give yerselves a break."

"We'll give it a think," replied Kerr, promising to meet him again.

Next day, when they were bolting down a transformer in the 'tween decks of the 'Oregon', the discrepancy between what could be earned by an electrician in New York and the weekly pay he was getting as a 'Sparks' in the Navy Yard, rankled in Kerr's mind.

"Can't get over that contractor an' his talk about big money last night, can you?" he asked Mark as he passed him a spanner.

"No," replied Mark. "Anyone in the electrical line is making big money in New York right now. But look at the way they've got to slave for it. I've seen 'em at it day an' night in the freezing weather, goin' without sleep an' meals to earn a few more bucks. Some of 'em become money-mad and lose health and everything to pile up the dollars.

"Anyway, we're tied to the Navy, aren't we?" he continued. "You've learned all you know in the Navy haven't you? Wouldn't think of quittin' for the sake of a few lousy dollars, would you?"

Kerr sensed the rebuke in his voice and glanced at Mark. In his dirty white overalls and officer's hat, with the Navy badge over the peak. Six feet tall, spare, with the dreamy eyes of a poet, he was an infinitely patient and wonderfully efficient officer.

"No, didn't quite mean that, Mark, but I'm beginning to think I'll never make a pile in the Navy," he added dolefully.

"Well, what if you don't, what if you never make a pile, maybe you'll be all the better for it," retorted Mark in a manner that was new and strange to Kerr.

"Look here, Kerr," he continued, "money's all right in its own way, but it isn't everything. Few of the people who've made millions became really happy, or even popular. Most of them are self-centred grabbers and are detested for that. It's only the power of the wealth they represent that makes others show them any respect at all. And, don't forget, many a millionaire would trade back all his wealth for the youth, health and happiness he lost while pilin' up the dollars. Whenever I hear guys talk like you, I think of my own brother George."

Up to this he had never mentioned that he was related to George Westinghouse, inventor of the world-famous pneumatic air-brake. Kerr listened with new and respectful attention to the officer's words:

"George was in the Navy like myself," he continued, "graduated at Union College, Schenectady. He was happy enough until he got the idea he'd become a great inventor, then nothing could hold him. If he had one, he had a dozen ideas worth patenting. Worked on 'em day an' night and spent all he had on them but could never raise enough money to see 'em through. One by one, other men hit on the same ideas and beat him to it, by getting the inventions on the market while he was still walking around with them in his head.

"Just when his health was on the break he got his air-brake idea, knew it was gilt-edged and it steadied him up. George worked on the development of it in all his spare hours, became obsessed by the invention and scrimped and saved his Service pay to get enough money together for the patent rights. When at last, he got the invention patented he went around from one manufacturer to another, trying to have a model made and tested on a railway car but failed all along the line.

"Well, George carried those Patent rights around in his pocket for years, worn out and half-crazed by the idea that someone would somehow forestall him again, as this sort of thing was happening every day with patents and inventors in the States.

"Then his chance came and he made the most of it. One day the Navy gave a banquet at Delmonico's in honour of old 'Commodore' Vanderbilt, who owned the New York Central railroad, and George was among the officers invited. With the connivance of the head waiter he got a seat next to the millionaire and while the dinner was in progress tried to get him interested in his air-brake.

"He explained how the device worked but the crotchety old plutocrat didn't want to be bothered. When George in desperation asked him would he have the idea tried out on one of his railway cars, the 'Commodore' flushed and replied gruffly, "Look here, young man, I've no time to waste on fools," so that was that.

"The dirty old meaney," Kerr interposed.

"Well, his luck turned at last," continued Mark, smiling. "'A couple of months later he was introduced to another railway owner, Henry Villard, who saw the possibilities of the invention at once. Mr. Villard placed a machine shop at his disposal and George got the engineers to make the first working model of the Westinghouse pneumatic brake and fit it on a railway car for testing under working conditions. As you know, it is now used by railway companies all over the world.

After hearing the story of his brother's years of heart-break, frustration and ultimate success, Kerr's regard for his senior officer zoomed to hero-worship. From that day on, he never allowed himself to be gulled by the lure of easy money, knowing that in the long run, money gained from hard work and creative ability is far more satisfying that profits derived from slap-dash piecework or other form of chancey industrial pursuit.

14

THE ELECTRIFICATION of the 'Oregon' and 'Massachusetts' was completed in the fall of the year and after running-in, the ships were commissioned for service. Mark Westinghouse was appointed Lieutenant Engineer on the 'Oregon' and would have taken Kerr with him as crew member, but the authorities decided that he should remain on the Navy Yard staff.

Kerr was promoted to Leading Electrician and sent to work on the 11,300 ton 'Iowa', a battleship which had been launched and was waiting in dock for engines. In the course of his electrical work, Kerr had many instructive and sometimes breath-taking views of the lifting, handling and placing in position, of huge engine and boiler sections by means of giant cranes and gantries, the assembly of heavy gun turrets, shell delivery machinery and other installations of special equipment that go into the construction of a modern battleship.

Working in the midst of this highly-skilled and superbly planned dockyard activity, he gained an insight into the way things were done on a grand scale and stored the knowledge in his mind, against the day he might be called upon to supervise similar operations as a practical engineer.

Kerr missed his friend and counseller Mark very much as the days passed by and especially during time-off at weekends. He had become absorbed by New York and was eager to explore. He revisited the places where a few years before he had been engaged in a fearsome struggle to survive. He wanted to obtain a revised picture of the living conditions in Manhattan and to try to analyse the social and political set-up that was to some extent responsible for the lives and destinies of its struggling millions.

He ascertained from public statistics that over three-quarters of the population of Manhattan Island was housed in tenements and that the boom in the construction of this type of building was at its peak in 1890.

Except for the cluster of massive public buildings and the huge monolithic skyscraper formation of the spectacular and much publicised New York skyline that stretched from Battery Park to the City Hall and included a few sections of Broadway, the rest of Manhattan was laid out in vast, well-defined slum areas each with a character and individuality of its own.

There were several grades of slums and of the people who inhabited them. Some of the sections were occupied by hard-working, law-abiding citizens who wouldn't have lived in them if there had been anywhere else to go. Others were inhabited by a type of people who would make slums in decent localities where they never existed before. There were sections over-run by gangsters and other criminal types and streets of notorious houses where the owners prospered on the profits from vice and crime.

The general standard of living varied in different slum areas but in all cases the people who lived in them degenerated morally, physically and mentally until they became as unfit for work as drones in a hive.

Most of the labourers and tradesmen who were building New York's skyscrapers, bridges, tunnels and public buildings, lived in a festering slum about two square miles in extent known as 'Irishtown' on the Lower East Side of Manhattan. North of this area were a dozen foreign colonies in which the inhabitants lived in an indescribable state of squalor, privation and despair. The middle-west section from Twentieth to Fiftieth Street was a vast slum area, where respectable working families lived under conditions that had steadily worsened for years.

When Kerr again walked the streets adjoining the Bowery, where the Jewish community had established the biggest ghetto in the world, conditions hadn't changed for the better. This section was known to the citizens as the 'Typhus Ward' and to the police as the 'Crooked Ward' on account of the number of crooks, thieves, fences and receivers of stolen goods who lived in its lanes and alleys.

This time he was in a position to view from a more impersonal standpoint the unrelieved miles of grim tenement walls with gaping windows where he had delivered cut-cloth to his clients in the sweat-shops. Apparently the fierce, overstrained tempo of life hadn't eased anything since he had worked there.

The whole district hummed and throbbed with sewing machines driven under pressure by human dynamos. Production still went on around the clock in rooms packed to suffocation point with bales of cloth and human beings.

Enslaved by the system upon which output was based, the sweatshops competed against one another to reduce the cost of labour still more and more and families worked themselves to death turning out the cheapest readymade clothing in the world. Finished coats, trousers, vests and underwear in sizes to suit all shapes and ages were made by the thousand dozen and were, as one producer put it "as alike as the feathers on a chicken's back."

Children of the ghetto swarmed around the tenement steps and every square foot of the sidewalks, and got in the way of traffic. Hundreds were killed or maimed every year.

Down below the approaches to the Brooklyn Bridge, between the prosperous business part of the city and the East River, Kerr discovered the worst section of slums in his travels. There were acres of them, one worse than the other. The tenements in this area were the most dilapidated he had seen. Many of the vast walls had no windows — the only source of air for ventilation came through filthy garbage shafts into which most of the slops and refuse was dumped. The district was occupied by hordes of tramps and rag pickers who existed with their families in misery and degradation. Children swarmed on the steps and in the halls, crawling about like ants. Three-quarters of the people in this section were Irish.

In damp unventilated cellars, polluted by sewerage from leaking drains as the tide in the East River rose and fell, hundreds of families lived amidst piles of rags, empty bottles, bales of used paper and heaps of stinking bones, collected for sale to the glue factories in the neighbourhood.

At night the cellars were alive with rats, human beings and scurrying bugs. A great number of them were stale beer dives, kept by women. The beer, nicknamed "razzle-dazzle," was composed of drink left in glasses or spilled on the counters of prosperous pubs; this was put into barrels and laced with wood alcohol. There was a thriving market for this liquor, which was the stock-in-trade of low-down drinking dens.

Tramps of both sexes were allowed to stay in these joints all night if they had the price of a round of drinks. If they hadn't the "entrance fee", the next best thing was to lie on the pavement adjoining a bakery or seek a spot to lie on, near the warm wall of a factory or brick oven.

Except in the Jewish quarter, saloons abounded, and they prospered in the slums. Social reformers claimed that fifty per cent. of the misfortunes of New York's poor were due to drink. The saloon was the one bright spot in a setting of utter misery and despair where the workmen, weary and disgusted with the squabbling and wretchedness at home, could obtain brief spells of bliss and forgetfulness in the company of his mates, or perhaps listen to some political Jack-in-the-box who claimed he could make the world a better place for everyone.

Women and children of the slums lived in an atmosphere of drink and became addicted to it early in life, to escape from the nightmare realities of their existence. The money squandered on tobacco and drink left little out of scanty wages for food, fuel and other necessities, and brought untold hardship and suffering into the lives of thousands of innocent victims.

A great source of social evil was the Black and Tan saloons off Kenmare, Schiff and Delancey Streets, on the fringe of the Bowery, so-called because they were frequented by the white and coloured races. Here the utterly depraved of both sexes met on equal terms and intermingled.

A feature of those saloons was the large barrel-type stove, set in the centre of the barroom that was never closed. At night, tramps and street walkers who had enough money to buy a round or two of drinks, were allowed to sit around the stove on a circle of chairs, but not to sleep. As soon as anyone nodded too many times or emitted a snore, he or she was manhandled and thrown out on the sidewalk by roustabouts, with the warning:

"Don't think your're gointa make a doss-house outa THIS saloon."

From the mingling of the races in this locality sprang a young population, who found themselves outcasts from the time they were babies. Instinctively, they formed into gangs and grew up in the world's toughest school, recognising only one code — the law of self-preservation.

Banding together against the police or any other force that tried to subdue them, these gangs formed the nucleus of a criminal fraternity that organised and expanded into a hoodlum empire, which a generation later overran and terrorised the inhabitants of many large cities in the United States.

When Kerr finished his circuit of the slums he had seen a living picture that was dreadful to behold or even to visualise. The voice of the social reformer was scarcely a whisper in 1891. When a man became unemployed through illness or other misfortune, he was pushed into the ranks of the destitute.

In those tumultous New York days, anyone in good health, able to work and pay his way was on the upgrade, with unlimited scope for enterprise and ambition. Kerr turned this over in his mind many a time, as he toiled away and increased his professional knowledge, in the shell of the battleship 'Iowa' in the Navy Yard, Brooklyn.

Kerr began to have notions, that he would like to get married, settle down and raise a family; but he worked it out to three decimal places that he could never save enough out of his meagre Service pay, to lay a proper foundation for this happy and desirable state.

There was abundant evidence of the fate of tens of thousands of young New Yorkers who fell in love, took a chance and got married, in the hope that better times were on the way. In a few years, nine out of ten found themselves with hopes shattered and a hungry family struggling for survival in a tenement.

Unwittingly, he was becoming more "Americanised" in his outlook on life. Up to now, girls were an almost unknown quantity to him, but be couldn't help noticing in the few contacts he had made, that there seemed to be little of the genuine love and affection about them that he had known in Ireland, and secretly longed for.

He was still living in Pat O'Driscoll's house and was introduced to plenty of girls at dances and other Irish social functions, but somehow couldn't hit it off with any of them. Girls seemed to become "Yanks" within three months of landing; then the man with the best job or the most money, usually claimed their attention — and got it.

Maureen and Una, Pat's pretty daughters were like sisters to him, and although he got many a hint and opportunity to get on more intimate terms, dread of the consequences of getting married put him off starting even a mild flirtation.

Hitherto his mind had been overshadowed by a sordidly depressing picture of life in New York, but there was another side to it of which he was becoming increasingly aware; a tantalising picture of success, luxury, wealth and fame which stimulated his ambition to strive for a worth-while place in the world. As he grew to maturity under the spell of New York, he imbibed its spirit and could hear its voice and interpret its moods.

Wealth, and the power that went with it, commanded respect from the mob. When appeals, hunger marches, riots and everything else failed, money could be relied upon to get things going, build factories, oil the wheels of production, grease the palms of corrupt officials, obtain big contracts or concessions, and if need be, alter the laws of the land.

Greed for wealth was the driving force behind most of the gigantic industrial and financial enterprises in New York. Organised charity was practically unknown there around 1900.

The social and commercial activities of millionaires and their families, kept the columnists of the great dailies supplied with interesting and spicy news.

In contrast to this there were countless intellectuals living and dying in poverty. But the public conscience was becoming concerned about the unequal distribution of wealth; the appalling hardship and suffering it caused, and the fact that there was no organised Christian charity to relieve the distress. Subscription balls were held in the millionaires' mansions on Fifth Avenue.

After every function, the morning newspapers carried full reports about the entertainment, the guests, the dresses, the famous gems that were worn and their value. The report was followed by a list of those who gave large sums and ended with the amount collected for charity. Guests who donated "princely gifts" came in for special praise, although in most cases, the contributions, however large, represented but a few grains from the vast store of their wealth.

As the subscription entertainments followed one another at intervals through the season, a blast of publicity was directed on them by the subsidised editors of an unctuous and servile press. The constant pluging of these laudatory reports gave readers the impression that the plutocrats had nothing but pity for the poor in their hearts, and that at last something was being done on a big scale to help the sick, homeless and unemployed.

The voice of the poor and oppressed gained strength when a great social reformer named Joshua Strong, addressed a meeting in the Bowery and said: "the greater the city, the greater are the riches of the rich and the poverty of the poor." A reporter of the Yellow Press fanned the smouldering fires of class hatred into a blaze, when he described New York Society as a self-exalted set, backed by leaders of State and religious bodies who placed themselves high above the ordinary working people.

The Yellow Press continued its campaign against the outrageous injustice of the capitalist system as it operated in America, and continued to publish articles which exposed the methods by which many of the plutocrats built up their millions.

The easiest, most popular and safest method of amassing wealth was to "squeeze a little more out of the poor." The poor and ignorant were so far gone in apathy and despair that they were unable to organise and hit back. The articles pointed out that the Astor family and established Churches, who between them, owned most of the vast slum tenement property in New York, were the greatest exponents of this scurrilous method of extortion, inasmuch as they got away with a heinous system of rack-renting for a whole generation, before they were exposed.

With the exception of the Police force, Fire department, Street car crews, scavengers, employees in the Water Supply and Drainage Departments and others directly employed by the "City Authorities," dire mass poverty was the lot of the general run of manual workers. Labourers worked eighteen hours a day for little more than sweat-shop workers.

The average wage for unskilled workers was from two dollars to four dollars fifty cents a week, on a time basis. Workers were fined for every minute of time they lost. The pay was so low in relation to the cost of living that no unskilled worker could afford more than one square meal a day. Anyone who protested about hours of work or rates of pay was instantly dismissed.

Girls in the great Department stores worked sixteen hours a day for one dollar and seventy-five cents a week. Waitresses worked the same hours for the same pay and had to buy their own aprons and pay the cost of breakages out of that. Child labour was always in demand as it was cheaper and plentiful, and helped to keep the pay of adults at a low level.

Although isolated labour disputes had occurred in many big cities in the United States, the "strike" as a weapon for backing up labour demands with forceful action had not been much used, as a "strike" could not be expected to last long when thousands of desperate people were waiting to jump into the first job that came their way.

As winter approached the plight of the unemployed rapidly deteriorated. It was the age of invention and new labour-saving devices were being introduced into almost every industry, creating more unemployment and making an already desperate situation worse.

Then, out of the lowest depths of apathy and despair, a resurgent spirit arose, and the trampled masses slowly revived and rallied to the call of the Socialist leaders. In 1884 the National Craft Unions of America had 700,000 members in its ranks. In the fall of 1887 nearly one million workers had joined the Knights of Labour, America's first recognised Trades Union, and by 1892 the total membership of trades unions in the United States had risen to two million.

In those eight formative years the long foreseen revival had materialised and grown strong. Those of the mob who had any fighting spirit left, responded to the call of Labour and before anyone quite realised it, the war against the injustices of the Capitalist system was on.

15

WHEN HE finished work on the electrification of the battleships 'Oregan' and 'Massachusetts' Kerr, now a Warrant Electrician, was put in charge of a dockyard torpedo assembly shop, under the command of a Shore-Captain, Nathanial Briggs, U. S. N. He had become expert in all classes of electrical work and was his senior's right-hand man.

One day he received a letter from home. The news was, that his brother Pat had emigrated to America and was working in a silver mine in Montana and that the old home in Borris-in-Ossory had been sold, as there was no one left to keep the farm going.

His mother had bought a house in Upper Gardiner Street, Dublin, and she and his sisters had gone to live there and run it as an hotel. The letter ended: "You must be nearly finished your time in the American Navy now. When will you be coming home? I'm getting old now and would like to see you again, before God calls me."

Kerr glanced at his watch. It was noon and he was due to attend at the paymaster's office for back-pay and Discharge papers in the afternoon. His seven years of service in the United States Navy ended that day.

A pay-off is always an occasion for a spree among seafaring people and Kerr wasn't one to break a tradition of that kind, so that evening he and another warrant electrician, Olaf Hansen, dressed up in "civvies", boarded a horse-omnibus that was about to cross the Brooklyn Bridge and after an hour's jostling and strap-hanging, got off at Madison Square to have another look at the sights of Broadway.

It was after dusk and the Great White Way began to blaze up like a comet. Incandescent gas lamps and sizzling gasoline flares were most conspicuous, but a good many shops were illuminated with the new "electric light." Some of the more go-ahead business houses had the latest and most fantastic advertising novelty of the time; clusters of coloured fairy lamps flicked on and off by means of a hand-switch. The whole front of a newspaper office in Times Square was lit up with carbon filament lamps set close together. This great wall of golden light shafted up into the sky, eclipsing all other illuminations in the neighbourhood.

Broadway itself, was at an early stage of development and its appearance was ever-changing. Massive sky-high buildings were humbled and depreciated by dingy, jerry-built shops and shanties, in their vicinity. The biggest department store in town looked cheap and out of place in its setting of fun palaces and huckster shops.

It was hard to make headway through the throng, slowly moving along the sidewalks of the world's greatest meeting-place. Unwashed and unshaven Irishmen, having a night out, in dirty boots and overalls, predominated.

Coal-black negroes were conspicuous by their colour. Red Indians in feathered head dresses, buckskin suits and beaded moccasins, padded through the crowd in single file, as if on a forest trail. Patriarchal Jews, in skull caps and long-skirted kaftans, moved sedately on their way. Chinese wearing pigtails and blue overalls, showed here and there in the sea of moving faces.

East Indians, dressed in turbans and flowing white robes, didn't look a bit self-conscious. There were Rumanians, Spaniards, Italians and Greeks. Women and children of undefinable nationality walked hand in hand or gazed into shop windows, in their thousands. It was a typical scene on a typical night on Broadway, the most cosmopolitan and congested thoroughfare in the world.

Dark, shabby side streets with an atmosphere of mystery and adventure, attracted the unwary but were really booby traps for drunks and greenhorns. Those who ventured into them at night ran a gauntlet of street walkers, pimps and pipe-benders (thugs who rolled a piece of lead pipe in a newspaper to use as a cosh), an easy prey to these parasites. Travellers were knocked out, robbed and often stripped of their clothing. Many of the victims died of exposure.

Broadway by night, was animated by the intensive activity characteristic of a giant ant hill. People were forever milling about, moving on towards some predetermined business of their own. Apart from the million or so sightseers on pleasure bent, there was an army of citizens working for a living and a horde of crooks, that found an easier way to live within its precincts.

All this ceaseless activity within a restricted area, created a volume of sound that had no equal anywhere and could best be described as a phenomenon, that started at dawn with the milk-carts and rose to a crescendo when the lights went on in the evening. It was a vast competition of noises: a horde of office cleaners working with scrubbing brushes and pails, an army of bakers and night-shift workers getting on with their jobs, caterers serving meals around the clock, bartenders, street sweepers, the clop, clop, clop of a thousand hooves on asphalt and the incessant roar of a million people talking at the same time.

Striving to make themselves heard above the din, were "barkers," shouting the attractions of waxwork exhibitions and freak shows, commissionaires, chanting the prices of seats in pit and stall, newsboys, yelling the latest headings and organ-grinders, trilling song hits from Tin Pan Alley or popular airs from an opera.

Olaf and Kerr, pushed their way through the swing doors of the Eldorado saloon and ordered schooners of beer. Bartenders worked like demons but couldn't keep up with the demand for drinks. The floor of the bar was paved with silver dollars. A bill poster on the wall indicated that Buffalo Bill Cody had a Wild West show somewhere up the street. The pair couldn't agree about going to see it.

Olaf said, "there's more to be seen where we are." They decided to stay put.

Conversation was free and easy in the bars and, as long as a man had plenty of money and threw it around, he was sure to have plenty of friends. It was a feature of life on Broadway that a man who had "struck it rich" would often walk into a saloon and call for "drinks on the house" until everyone got drunk and drank themselves sober again.

All classes and creeds fraternised in the Broadway bars: millionaires, playboys, notorious characters, politicians, confidence men and bums. There were no social barriers as long as one could pay for the drinks. In his rip-roaring days John L. Sullivan was the toast of Broadway, and could drink them all under the table. Although John L. had lost the championship to Jim Corbett, they were still talking about his marathon seventy-five-round knuckle fight in which he knocked out Jake Kilrain after a battle that lasted for over two and a quarter hours.

Nobody knew the Broadway saloons better than John L., who earned over a million dollars with his fists and spent most of them on drink. When he was flat-broke and reduced to borrowing, his friends gave him the go-by. This gave him a turn against drink and he became a powerful advocate for the introduction of Prohibition. The story of how John L. Sullivan rose again after his downfall and regained his self-respect and the respect of others, was an epic that made him a national hero.

After a few more beers Olaf and Kerr forgot all about going to a show and did a tour of some of the famous Broadway bars instead. Every saloon they visited, was packed with men and women singing themselves hoarse and behaving madly as if a carnival was on. The roisterous renderings of "Little Annie Rooney", "Bill Bailey", the "Lily of Laguna" and "Mollie Malone"; the latest Irish song-hit were no mean contributions to the sounds of Broadway.

In contrast to the poverty-stricken atmosphere of the bars on the East side, there was an abundance of everything on Broadway. Free soup, hot-dog sandwiches and all kinds of savoury snacks were there for the asking, provided one was buying drinks. Starving human wrecks came in to beg, but were ignored or ridiculed by the customers and ordered to "beat it" and hustled out by roustabouts. No one had any time or sympathy for the poor.

Weaving in and out through the merry-makers, were flyboys plying their sinister trade. Cocaine was on sale at a quarter dollar a pinch; opium cigarettes could be had a a few cents apiece. Furtive agents from the underworld offered gold watches and valuable jewellery for half-nothing,but Kerr and Olaf were too well-up to have any dealings with them, knowing too well that anyone who bought them would, as likely as not, be beaten up and robbed on the way home.

The night passed quickly and dawn was breaking when Kerr and Olaf parted at the Fulton Street end of the Brooklyn Bridge. Kerr walked home, a bit muzzy after the beer, humming his favourite song "Way down upon the Swanee River." He reached Pat O'Driscoll's house, slipped in quietly with his latch-key and was soon in bed, sleeping off the effects of his night out.

Next morning, he awoke with a sore head and couldn't decide what to do. He was now a free man with all the world before him, but, strange to say, did not feel one bit elated. There were plenty of positions with good prospects waiting to be filled, and he had promised to see a friend, John Holland, who was destined to become inventor of a submarine as soon as he left the Navy. Prospects for the future couldn't be better, but whether it was from the after effects of drink or just a reaction from his sudden break with naval routine, he looked on the world this morning with a jaundiced eye.

The United States was a marvellous country, he reflected; but what had it to show? In an introspective mood the snap impressions registered during his years in New York passed before his mind in cineramic review. The country was governed by an oligarchy of millionaires, who made laws to suit themselves and became rich by exploiting the labour of the poor. Organised Society was controlled by a set of self-opinionated hypocrites, who expected everyone else to knuckle down and live according to the parable in the Scriptures about "the crumbs from the rich man's table."

Religion in America was represented by an association of people who preached Christianity, but practised extortion of a most despicable kind. Religious leaders of all denominations, favoured private ownership and the protection of private property. No one seemed a bit concerned because the poor had no property and no protection against exploitation by the rich.

Civic administration in New York was maintained by a strong-arm police force that gathered revenue from the biggest vice organisation in the world. It employed criminals to do its dirty work, extort 'Protection money' from owners of notorious houses, gambling dens and honest business folk alike, it wiped out anyone who refused to toe the line, or who tried to expose its soul-destroying rackets.

The more he thought these things over, the more Kerr felt he could never settle down and live in America. None of the dazzling artificialities the country could offer; a brilliant career, the opportunities to make millions or anything else, would ever compensate for the genuine love andhomelinesshe knew could only be found in Ireland. The letter from his mother decided the issue.

His mind made up, he jumped out of bed and dressed and that afternoon booked a passage on the Red Star liner 'Lapland', which was sailing from New York to Queenstown at the end of the week.

Kerr arrived in Dublin in the fall of 1893, and went to the house in Upper Gardiner Street, where his mother and sisters had moved to, from the country. Their new home was a stone's throw from the corner of Mountjoy Square. No returned exile ever had such a welcome. His mother was overjoyed and proud, his sisters said he was "any girl's fancy." When the fuss of arrival died down after a few days, Kerr set out to explore the neighbourhood and get his bearings.

The first morning Kerr set out for a walk in tall silk hat, frock coat and the glossiest boots in town, he created quite a sensation. Sunburnt and handsome, he had the quick springy step of an athlete.

Who was he? What was he? The window curtains moved surreptitiously and the queries buzzed around. Then the answer, "a young engineer home from America," circulated. Before the day was out the girls in the neighbourhood had his pedigree and he was a marked man.

Around the middle of the century, Upper Gardiner Street and its environs, was the domain of fashionable Anglo-Irish society. The neighbourhood never lost its distinctive Georgian character. Some of its houses had been occupied by members of the notorious Hellfire Club; a band of young bloods whose dissolute ways, drunken brawls and duels, were perpetuated by fireside stories that still circulated like the unlaid ghosts of a lingering past. High Society, had dispersed and moved to more spacious houses in St. Stephen's Green and Merrion Square on the south side of the city, and the social status of the vacated areas on the north side began to decline.

Although Upper Gardiner Street had seen better days, it was still a genteel neighbourhood and professional men and their families resided in most of its houses.

The old Georgian houses with their granite steps, massive hall-doors, ornamental fanlights and wrought iron window balconies, still retained an air of solid respectability. All had deep basement areas guarded by heavy iron railings. Torch extinguishers, like giant cone-shaped candle snuffers, still remained in their places over fanlights or on high corner-posts of the area railings. In bygone days the torches, made from tow dipped in pitch, were held aloft by link-boys to light the way for sedan chairs or well-to-do pedestrians, before street lighting came into use.

Gas lamps lined the pavements of the secondary streets and old-fashioned sizzling electric arc lamps, similar to those used to illuminate the Avenue de l'Opéra in Paris in 1887, were being erected in Sackville Street by the Dublin Corporation. Main thoroughfares were paved with uneven stone setts, and road sections outside churches and schools were surfaced with hardwood blocks to deaden the noise of traffic.

Old "Dublin Southern District" horse-trams moved on rails in the centre of the road, and pulled up anywhere on the route to pick up or set down passengers. In comparison to the speed of traffic in New York, everything in Dublin seemed to move at a crawl.

This thought was running through his mind as Kerr walked to Nelson's Pillar, the centre of the business part of the city, in about ten minutes. To all appearances, time didn't seem to count for anything. Trying to adjust himself somewhat to the leisurely tempo of his new environment, he went into a pub and called for a bottle of stout. He wasn't long there before he realised, that one could learn more about the city in an hour from pub-talk, than from studying guide books or asking questions for a week. The alcoholic vapourings of idlers and know-alls, issued the topical news in tabloid form as good as any reporter's notes.

"Belle o' New York's showin' in the Gaiety," said a loud kind of voice.

"Yeh, an' showin' more'n what some o' th' holy men think should be shown," remarked a man with a pint. A ripple of tipsy laughter followed. With the unpredictable meanderings of beery inspiration, the conversation suddenly switched to public transport systems and trams and Kerr listened intently.

"Nearly bein' a right accident at the bottom of Constitution Hill yesterday," said someone, by way of starting something.

"There was a relief pair o' horses hitched on to one of the trams an' when they got to the top o' th' hill at Christchurch Place, two of 'em slipped on th' frozen setts, an' th' tram backed down th' hill gainin' speed all th'time. Such a panic! Women an' kids screamed, men jumped off an' th' horses kicked an' lashed out in all directions. One of 'em broke his leg an' had to be shot."

"A wonder no one was killed," someone remarked.

"Them horse-trams is outa date," said another in a voice of muzzy superiority.

"Aint th' whole bloody country outa date? A city like this with horse-trams! There's electric trams everywhere now," chimed in another to keep the conversation going.

"Th 'lectric trams is comin' but there's money in 'em an' there's a tussle between th' Dublin Corporation an' William Martin Murphy, about who's gointa get th' rights o' way to put up th' wires an' lay down th' tracks, there's laws about all them things," said a man with a newspaper, trying to impress everyone with his knowledge.

"Arrah, don'tya know the Corporation wants to run the trams an' make money outa them, but they haven't an engineer who could tackle a big job like that. There's no one in this country who could start an electric tramway, except William Martin Murphy an' you'll see th' Corporation'll have to climb down an' give him the rights o' way." Later events proved these words to be prophetic.

Immensely interested, Kerr took everything in, without joining in the conversation. His mind, buzzing with the news, he finished his drink and started for home.

Margaret, Kerr's mother didn't take kindly to her new environment. The congested atmosphere of the city was so different from the open-air spaciousness of the farm at Bawngorra, that sometimes she found it difficult to breathe and finally settled down in a high-backed chair to wilt and fail, like a transplanted tree that couldn't take root in new ground.

During the Christmas season, there was scarcely a night when the drawing-room windows on the ground and first floors of the houses in Gardiner Street, Temple Street and Mountjoy Square, were not open to the full, emitting streams of light and merriment into the gloom. The music of piano and violin, young voices uplifted in song, shouts of laughter at the game of 'Musical Chairs', the swish of silk as partners swung at the 'Lancers' or the thump, thump — crash of feet when everyone joined in a 'Sir Roger de Coverly.'

At one of these parties, Kerr met a girl named Beatrice, fell in love with her and after a short courtship they decided to get married. The days passed in a delightful whirl of preparations for the wedding. The custom of the bride making her own trousseau was abandoned, as it would take too much time, so the pair started on a shopping spree to buy everything ready-made. There were visits to the dressmaker and the jeweller, introductions to friends and relatives, engagement parties, visits to the theatre and bachelor parties. Kerr spent money as if he owned a gold mine. Both of them lived in the clouds and couldn't even get down to deciding whether they should emigrate to America or try to settle down and live in Ireland.

They were married in the pro-Cathedral in Marlboro Street, inundated with confetti, blessings, presents and good wishes from a crowd of friends and relatives.

When they arrived home after the wedding breakfast, a letter awaited Kerr from a firm of bacon-curers in Limerick. It was in connection with an advertisement for the services of an electrical engineer to which Kerr had replied a month before and forgotten about. The letter contained an invitation from Mr. Shaw to visit him in the capacity of consultant, to advise him in the matter of changing his factory from old-fashioned, hand-operated methods of production, to modern electrically operated equipment.

The happy pair decided to be practical and spend their honeymoon in Limerick, with the prospect of Kerr taking up a post there later. They just managed to catch the afternoon train to Limerick, and booked into an hotel there on arrival.

16

AFTER A wonderful honeymoon during which they explored the city, visited King John's castle and other romantic places on the banks of River Shannon, the day arrived when Kerr had to keep his business appointment.

Mr. Shaw, a handsome middle-aged man received him cordially and made him feel at ease. He was interested in industrial progress and especially in electricity and its application to machinery. They discussed his bacon-curing and carcass processing plant and its output, in detail. He took an instant liking to Kerr.

"We're working here with worn out, obsolete appliances, and my aim is to build the most up-to-date factory in Ireland," continued Mr. Shaw, with the outspokenness and assurance of one to whom money doesn't matter.

"Name your salary, Kerr and you'll have a free hand with everything; we'll plan the whole thing together," he said, convinced that he had met the right man.

The sudden question about salary stumped Kerr. There was no "rate" for electrical engineers in those days, as the profession was not properly established and one of Kerr's weak points was to underestimate his own value.

On the spur of the moment he named a salary which was about half what Mr. Shaw thought he would expect. Mr. Shaw, shrewd business man that he was, never blinked an eyelid, but just expressed delight at having secured the services of a competent engineer. With a mutual feeling of confidence and friendship they shook hands and the interview ended.

Kerr returned to the hotel, elated at having secured steady employment for the next couple of years, as his savings were running low. Beatrice was delighted at his success. As it was beyond their means to continue living in an hotel they went house-hunting next day, and ended up by renting a small house in Davis Street.

A week later furniture and household effects, including a beautiful piano, arrived, after a five-day voyage on the canal boat from Dublin. They were not long settling in their new home and a period of great happiness followed.

The neighbourhood reminded them of Dublin, surrounded as it was by streets of Georgian houses with the same distinctive features of deep areas, iron railings, steps up to the front doors, ornate fanlights and iron balconies in front of the windows, and brickwork mellowed to a sickly yellow, just as if the houses had been transplanted en bloc, from Dublin's Dorset Street or Merrion Square.

In 1894 Limerick city had every right to be called "ancient." There were no gaslights in the streets or houses, no telephones, no motor cars or public transport of any kind and the luxuries of baths and lavatories were enjoyed only by a privileged few. A very low standard of living prevailed among the working classes and Samuel Lover's "stage Irishman", was a living reality.

The Royal Irish Constabulary maintained law and order in the County and the City Corporation employed Watchmen, to arrest footpads and thieves who infested the streets after dark. Watchmen patrolled the streets at night, swinging rattles, calling out the hours and generally making a racket to let everyone know they were on duty, so they could sleep in safety and comfort. They were tolerated by the ratepayers as a necessary evil. The sidewalks in the main streets were paved with flat stones but in the side streets they were muddy and criss-crossed with wheel tracks.

The city itself was a stronghold of class distinction, vested interests, bigotry and snobbery. There was an affluent society of Anglo-Irish landed gentry, Army and Naval officers, a sprinkling of snobbish, high-up Civil Servants and a quota of parasites who supported the régime and prospered under its protection.

British rule and all its works and pomps was detested by the vast majority of the townspeople, who regarded it rightly or wrongly as the direct cause of their poverty and heritage of political woe. Irish people who sided with the British or fraternised with them in any way, were regarded as traitors to their country and were ostracised by their own kith and kin.

Kerr was not long resident in Limerick before he found himself a kind of social misfit. His education and professional status placed him on the same social plane as a senior Army officer and he would have had no difficulty in being proposed and accepted as a member of any of the city clubs. However, as an Irish-American, he had no wish to be seen hob-nobbing with those who upheld British rule and the only alternative was to mix with nationalist fanatics who were forever making preparations for militant action of one kind or another. This was not to his liking either, so he finally decided to steer a course between both factions, settle down to the job in hand and leave politics to the politicians and fighting men.

After his appointment, he turned the front parlour of their new home in Davis Street into an office, got out his drawing board and tee square and began working on the plans for reconstruction and electrification of the old factory in Mulgrave Street, which at that time ran into a field, about two miles from the centre of the city.

The original factory was a sprawling collection of tin-roofed sheds and outhouses where over a hundred men and women were employed in the killing room, curing bacon and hams and manufacturing other pork products, for which the firm already had a big market in Ireland, England and America. Practically all the work was done by antiquated hand methods.

Kerr could never get over the lack of transport facilities and bad roads in Limerick. Used to the clean sidewalks and acres of asphalt in New York, he abhorred dirt and mud. With the inbred naval habit of keeping a smart appearance and maintaining the dignity of his profession, he set out to walk the distance from Davis Street to Mulgrave Street each morning in top hat, frock coat and polished boots, but before he got very far, his boots and trousers were as muddy as a ploughman's. The alternative was to accept a lift on an ass and cart of which he had plenty of offers and tempted as he often was to take one, he felt that if he did, his incongruous appearance would make him the laughing stock of the town.

When the people got to know that he was the first electrical engineer ever to come to Limerick, they treated him with deference almost amounting to awe.

"Look, there's Mr. Quaney that's goin' to bring th' 'lectricity here" or "Aw, he's another Edison," they'd say, trundling along on their asses and carts. When he went around the old factory it was all "Yes sir" and "No sir" or "What can I do for you, sir?" Everyone saluted and salaamed and did their best to please him.

The first plan off his drawing board was for the new boiler house and hundred foot chimney stack, which stands as a local landmark to the present day. A contractor was employed to build the chimney and boiler house also to extend the roof in the curing house. Mr. Shaw worked with Kerr on the master plan. Kerr estimated the power requirements and ordered the machinery. Mr. Shaw kept the accounts and never questioned the cost of anything.

"If you say it's the best equipment for the job we'll get it, Kerr," he would say, reaching for his cheque book and writing out the order. Mr. Shaw supervised the construction of the new buildings and extensions and Kerr got out the blueprints for the machinery, shafting and belt drives, and ordered the electrical power units to be used for the manufacture of brawn and sausages.

There was a burst of activity seldom seen in those parts when the three six-foot diameter Lancashire boilers arrived by canal boat. The heavy sections were loaded on drays and hauled by teams of horses and men, over shockingly bad roads, to be assembled in the new boiler house. Fitters arrived from England to assemble the boilers.

Then stories began to circulate about the pressure they would generate when the furnaces got going. It was an era before safety valves were perfected and lurid tales had come through in recent years about boilers bursting in London, Glasgow, Newcastle and other great industrial cities, and the number of people who had been killed or maimed by explosions. Most of the men were afraid to enter the boiler house, or to have anything to do with the boilers and gloomily predicted that local men would never take jobs as stokers.

Kerr worried a lot about the time he had to waste walking to and fro from work although really it helped to keep him fit. His young wife was often lonely and depressed when he didn't come home for meals. Neither of them took much interest in local affairs and few friends called to pass the time of day. Beatrice whiled away the hours practising at the piano and developing the wonderful talent she had for music.

On Sundays Kerr would listen to her playing, enraptured, then suddenly wake up and start planning furiously for some way to amass enough money to buy a house and build up some security for the future.

After months of work, the new boiler house was completed and the three boilers were installed as planned. The great hundred-foot brick chimney reared up, forming an appropriate monument to industrial progress. Two 'Lightfoot' steam engines with six-foot diameter flywheels, arrived in sections, were assembled on the engine beds and connected to the boilers. Then four 'Z' type dynamos arrived from London via Dublin by rail.

The dynamos were of the same pattern as those used by Edison to provide electric light on the S. S. 'Colombo' which in 1880 sailed on an epic voyage from New Jersey, around Cape Horn to San Francisco — the first liner ever to be equipped with electric lights. The dynamos generated for 115 carbonised paper filament lamps and not a single bulb failed during the voyage, which lasted for two months.

The task of estimating and planning had now reached its peak, and all that remained was to put the plans into practice. Mr. Shaw and Kerr were beginning to feel the strain of continuous overwork. At this stage one of their chief worries was the shortage of skilled technicians to do the fitting and pipe work.

Another thing was, that their overmastering desire to fulfil an ambition had not inspired the local workmen to greater efforts. Naturally indolent and sublimely ignorant, with a come-day-go-day outlook, the countrymen had little interest in any sort of progress and a profound distrust of machinery, regarding it as an enemy of their class that would cause more and more unemployment.

Although patriots sang "The West's Awake," with great gusto, it was still in a trance industrially. Labourers started work in their own leisurely way and made a show of effort until someone produced a pipe and plug of tobacco, then everyone within talking distance followed suit and took it easy for a spell. If a stranger came near where men were working, they stopped automatically to find out who he might be, and what was his business and until these important questions were settled no one would think of starting work again. Among the working class time didn't matter and the word "hurry" was taboo.

Trained to work in an environment of American drive and efficiency, the complete lack of interest shown in what was going on, annoyed and disheartened Kerr. If there were two ways of doing a job, it was invariably done the wrong way and the supervision of handling and assembling intricate machinery was, at times, a penance.

When the dynamos and cables arrived an unforeseen problem cropped up. An arrangement had been made for the manufacturers to send electricians to instal them but a letter arrived, with the news that no electricians were available and would not be, for some months.

The possibility of a long delay at this critical stage of the work worried Mr. Shaw and he discussed the matter with Kerr to find a way to avoid it. There wasn't an electrician of any kind in Limerick and it seemed unlikely that a satisfactory man would be found even if they searched the whole country. Kerr felt somehow that the onus was on him to see the job through and the only way to do it was to tackle it himself, a risky task to undertake with the type of help available, but this was a challenge to his pride and professional integrity.

Beatrice was shocked when Kerr returned home and told her of his decision. Brought up with strait-laced ideas about the ethics of "class" and all its delineations, she couldn't understand how "he could make himself so cheap."

"But Kerr, you're a qualified engineer, surely they can't expect you to go back to being an ordinary workman?" she enquired fretfully.

"There's no other way, dear. I couldn't walk off and leave the job as it is, could I?" he replied quietly, and so the matter was settled.

When Kerr decided to pocket his pride and work to see the job through, he didn't fully appreciate the disparagement he was letting himself in for. It was like having to face the world after being degraded in rank. When he was going around with a sheaf of plans, marking out the places for machinery and teaching workmen how to read blueprints, everyone looked up to him. He was a "somebody" able to make his living with clean hands, the envy of the artisans; but the fall of Lucifer was no greater than his, the day he turned up in a boiler suit and heavy boots, carrying a tool bag.

"Well, there's a come-down if ever there was ... an ordinary workman ... an' we thought he was a gintleman," they whispered. People kept out of his way as if he was a leper and he had some difficulty getting labourers to help him.

"Always thought there was somethin' queer about that fella," they went on. "If he'd joined a club an' mixed around with the gintry as he should, he'd a bin a Knight or somethin' by now," was their verdict.

But Kerr, knowing the world as he did, wasn't too perturbed about the psychological effects of having "lost face," although he did feel a bit deflated one morning when one of his helpers answered respectfully "Yes sir" and the other blurted out:

"Arrah, what are ya sirrin' him for, for Jasus' sake, hasn't he to work the same as ourselves?"

More than once he heard himself described as a "Muggadhaun", but notwithstanding the slurs and innuendoes, he got on with the job and his interest in it neutralised the hostile attitude of the rank and file.

When Beatrice opened the hall-door at dinner-time and saw him in dirty overalls, begrimed with toil, she almost burst into tears. She hadn't seen much of the ups and downs in life and felt humiliated.

However, young in heart and outlook, they soon adjusted themselves to the new conditions. Such trifling problems were swept away when a son was born. The tiny infant brought more love and interest into their lives and time passed quickly as they planned for the future.

Then one morning Kerr came across an article in the local newspaper that revived old memories and rekindled his ambition to become a great inventor. While the breakfast was being cooked he read it out to Beatrice: "Electrical engineers have as yet been unable to perfect a system of working tramways electrically along crowded thoroughfares. The overhead trolley system is being continually found fault with in America. Inventors have long been engaged endeavouring to overcome the difficulties and as an incentive to them to throw themselves into their task with renewed vigour, the Metropolitan Traction Company of New York has offered the handsome reward of £10,000 to the enterprising inventor bringing out a system of street car propulsion which will be superior or equal to the overhead system, but without possessing the objectionable features of the trolley, for crowded thoroughfares. The idea is stated to be, to encourage some sort of underground trolley system for street cars."

"Now there's something in your line worth going in for," commented Beatrice enthusiastically. Once again Kerr felt the call of America but this time he reflected a little sadly that he could not answer it, after taking on the responsibility of a wife and family and the moral obligation to build a home and provide for their future.

The reading of that report unsettled Kerr and stimulated him to finish his contract, so that he would be free to start on another assignment at a higher salary. The rebuilding of the main factory was almost completed. A new ammonia plant and installation of pickling vats were ready for service.

Under normal conditions three or four electricians would have been required to do the work that Kerr tackled single-handed. It was a vast undertaking. He hadn't one reliable helper and the installation of wiring and testing of power and lighting circuits, was slow and arduous work. When the day came to connect up the steam engines and dynamos that were to generate electricity, he was nearly "all in."

Early one morning in March, 1896, everything was ready and after a final check up of the electric wiring Kerr gave the mechanics the "all clear" to stoke up the furnaces and carry out a test.

Clouds of smoke arose from the main chimney, the roar of the furnace increased in volume and a couple of hours later, the steam gauges registered sufficient pressure for a test.

After a series of valve adjustments, spurts of water and escaping steam, the great flywheels began to turn slowly and gain momentum. The hum of the dynamos rose to a high-pitched whine and clusters of sparks, flickered around the armature contacts as the generation of electricity began.

It was dark that evening when Kerr went the rounds, putting on the light switches and running the belt-driven machines until the whole factory blazed with the new electric light and throbbed with steam and electrical energy.

Thus under the direction of one man the miracle was accomplished and the first factory in the west of Ireland to be run by electricity was ready for service. There was no fanfare of trumpets or gathering of civic dignitaries to usher in this historic event and the crowd of factory hands and rustics that gathered to watch the switch-over could not grasp the full importance of the occasion and dispersed when the lights went out no more impressed than if they had seen an annual Guy Fawkes Day display.

But it was a thrill-packed day for Kerr and Mr. Shaw the two partners in the drama. An unusual silence fell between them when the strain of waiting and watching eased. They were too overcome with relief and pride of achievement to do anything more than shake hands and express a fervent thought "Well, that's over at last, thank God." Mr. Shaw knew that the success of the whole undertaking was due to Kerr's efforts and respected him all the more, because he was the most efficient and unassuming young man he'd ever met.

Owing to popular prejudice against the introduction of steam, some time elapsed before anyone would come forward to learn to stoke the boilers, but eventually a few men who had acquired experience in a saw-mill, attracted by the high wages, offered their services.

As soon as the improver mechanics became familiar with the running of the plant, the fly-wheels started to hum again and with the help of Kerr's advice and training, the men applied themselves more efficiently to their duties, until the pulley-belts whirred with greater velocity and the factory, like a fully synchronised power unit, increased output steadily day by day.

Although the steam-driven section of the plant was now running smoothly the same could not be said about the electrical department. Few of the workmen could be induced to have anything to do with electricity. They had an instinctive fear of it. Electricity had the power to shock and kill, like lightning.

The youngsters who attended the machines would feed them with meat all day long but refused to touch a switch, even to turn the power on or off, so an advertisement was put in the newspapers for a maintenance electrician.

Now that all the manual work of assembling boilers, dynamos and shafting was done, Kerr resumed his original status as an engineer, debonair, efficient, advising, directing and helping those, whose interest in the new machinery was at last awakened. Acknowledged by all as a master technician, he was the authority on everything connected with the machinery.

Within a few months of the switch-over, the factory output statistics showed a remarkable rise, and soon the firm's products became known for excellence and quality in many parts of the world.

After the long period of strain and overwork, Kerr enjoyed a spell of rest and relaxation. The prosperity of the factory was soon reflected by increased earnings of the workers and Kerr became a leading figure in the town and the hero of every school boy.

Now that his contract to modernise the factory was to all intents and purposes finished, the unspoken question arose whether Kerr would stay on at his post of engineer, or leave it for some other field of service. With that delicacy of feeling between true friends, neither liked the idea of parting or even broaching the subject. Although Mr. Shaw gave many a hint that Kerr's post was permanent if he wished to stay at it, no further agreement was entered into, and Kerr left the matter at that. Prompted by ambition he already felt fit to take on and grapple with another engineering problem.

In August, 1896, a letter arrived from Dublin that opened up new vistas and thrilled the young couple with the spirit of adventure. It was from Kerr's sister, with the news, that at last, after years of talking, an Electric Tramways Company was being formed which had gained permission to provide a public transport system for the whole city of Dublin. There would be wonderful opportunities for engineers who were "in on the start." Kerr and his wife talked things over and decided that it would be in their best interests, to move back to Dublin.

The following week Kerr interviewed an electrician who had applied for a maintenance job in the factory. After satisfying himself that the man was competent enough to look after the electrical equipment, he told Mr. Shaw of his wish for a change. With a mutual feeling of regret the friends parted, never to meet again.

17

DUBLIN IN 1896 was coming to the end of a long period of smug Victorian rule. Although sedan chairs were out-dated and relegated to scrap yards, the city centre around Sackville Street still presented a picture of unhurried tranquility. Plenty of men were still to be seen on horse-back, rich people travelled luxuriously in carriages and landaus, with top-hatted and liveried footmen, and the penny farthing bicycle was becoming popular although regarded by some as a dangerous nuisance. Those who could afford it, travelled on outside cars, driven by jarveys in fancy dress and light grey bowler hats. Mounted policemen in blue uniforms and spiked helmets cantered here and there on well-groomed horses, to control crowds.

Senior civil servants, aldermen, doctors, lawyers and other professional men, wore striped trousers, frock coats and tall shiny silk hats. Spats were sported by racehorse owners, theatrical celebrities and bankers. In the afternoons Grafton Street was thronged with beautiful women, dressed in the latest fashion.

In the evenings, British troops in colourful uniforms, dragoons, hussars, guardsmen and kilted Highlanders, swaggered along the pavements from the Rotunda to

Nelson's Pillar and on to Dame Street. On Saturday afternoons the air was full of music. The Boer War was brewing and military bands blared out:

"Ta-ra-ra- Bom-Bom-Bom-Bom-bay,
Ta-ra-ra- Bom-Bom-Bom-Bom-bay,"

and

"Good-bye Dolly I must leave you."

In the back parlours of dingy pubs, Irish patriots nursed their spleen, plotted rebellion and sang:

"Down with the Queen and her ould tambourine
And Hurrah for Kruger's army."

The workpeople of Dublin, were desperately poor and the slums in which they lived had the name of being the worst in the world. Workmen went about in scare-crow clothes and the women wore long, heavy shawls over their heads like cloaks. Nearly all the children of the poor ran about in their bare feet, in summer heat and bitter winter sleet. Workpeople walked to and fro from work; those who could afford to travelled on the few horse-trams that were available.

But the city itself was awakening to a sense of its growing importance as the capital of Ireland. Railways had been opened up throughout the country. Industries slowly developing through the years, prospered and expanded. Shipping interests; booming all over the world, were forming trade links between the port of Dublin and far-flung outposts of the British Empire. Dublin's Customs House became a clearing house for British industries and a gold mine for Her Majesty's Treasury.

As the city developed, its population increased and the upsurge of industrial activity created an urgent need for a new and modern public transport service to cope with the ever-increasing tide of thousands of citizens going to and from work, or on an annual exodus to the seaside.

William Martin Murphy was the first great Irishman to sense the urgency of the situation, and attempt to do something practical about it. He was without doubt a financial genius of the Vanderbilt type, able to think years ahead, gain the confidence of the people, gather in vast amounts of borrowed capital and invest it in approved projects with the sure knowledge that it would produce big profits in a few years' time.

Martin Murphy had considerable experience of public transport undertakings. He rose to become a director of the Dublin Central Tramways Company, which had successfully operated a line of horse-drawn trams within a short radius of the city centre since 1886.

There were several other contemporary financiers who had visions of becoming transport magnates by exploiting the needs of Dublin's citizens, but had neither the "go" nor the backing, to outwit this master-mind in the struggle to transmute their dreams into dividends.

The attitude of the Dublin Corporation and the British Houses of Parliament in London, was a big handicap to the establishment of a satisfactory public transport service in Dublin.

If a company wanted to open up a line for public convenience, permission had to be obtained from the municipal authorities to lay down the tracks. Year after year, the Dublin Corporation dilly-dallied with applications, or promised to "go into the matter," hoping that as time went on and the need for transport became greater and the prospect of quick profits became rosier, companies would come forward and offer bigger rents for select and potentially profitable routes.

If approval was obtained from the Corporation, a Bill had to be prepared and sent to the Houses of Parliament. If the Bill was passed by the House of Commons, it had to go on to the House of Lords and if the House of Lords approved and passed it, the Bill was placed on the Statute Book and became Law.

After many years of haggling and frustration, a transit enterprise known as the North Dublin Street Tramways Company, obtained permission to open a line of horse-drawn trams from College Green to Inchicore on the south side of Dublin, but when the Company was about to start laying down tracks, some of the promoters became faint-hearted and withdrew their support.

By purchasing a large number of shares at a very low price, William Martin Murphy kept the Company floating and work on the new line started in 1876. The line prospered, and after a few years Murphy had gained a controlling interest in it.

In 1881 the Dublin Central Tramways Company agreed to a merger and the Dublin United Tramways Company was formed, with William Martin Murphy acting as advisor to the Board of Directors. This Company now represented most of the financiers interested in Dublin's public transport, and with competition virtually eliminated, it had no difficulty in obtaining way-leaves from the Dublin Corporation. When it came to having Bills approved by Parliament, the weight of influence and invested money turned the balance in its favour and practically all the subsequent applications for rights of way, made by the D. U. T. Co., were passed by both Houses, unopposed.

During the following years, the D. U. T. Co. consilidated its position and opened up many new tram routes. 1885 was an eventful year, in which a new line was put into service and an individual, who was destined to become one of the most colourful characters in the history of Dublin's public transport services, made his début.

He was a young man in his twenties, whose mates knew him as an independent sort of character, who had a furious temper when roused. His name was Cliften Robinson and he was employed as a tram driver, on the Nelson's Pillar — Sandymount line.

Near the Sandymount end of the line, there was an old hand-operated swivel bridge that was used to allow coal boats in and out of Ringsend dock, with cargoes for the Dublin Gas Company. When the bridge was open, the trams were delayed and that happened at least once a day.

One afternoon when Cliften's tram stopped at Westland Row railway station on its way to the Pillar, an irate inspector bawled him out for being late. Cliften's reply was to jump down off his driving seat and challenge him to fight. He was sacked on the spot. Cliften emigrated to America and many years elapsed before he returned to Dublin, to re-enter the public transport service arena in a new and dramatic rôle.

Between the years 1881 and 1894, the D. U. T. Co., consolidated and expanded its business, branching out into new lines wherever it seemed likely they would pay. In 1881 the Company carried 10,000,000 passengers and thirteen years later 20,000,000 people were carried at a flat-rate of one penny per head. The average fare was "A penny to Nelson's Pillar."

During the year 1894 however, it became clear that a large section of the community was still not catered for and many expressed the opinion, that horse trams were out-of-date and responsible for slowing up the whole business of the city.

However, as there was no competitor service, it didn't seem that anything could be done about it. The D. U. T. Co., now, had complete control of the public transport service in Dublin and was happy enough, taking in the cash and paying substantial dividends to its preference shareholders. The Company was proud of its hundred and eighty-three horse trams, plying the streets and had received permission to open up a new line from Glasnevin to Dolphin's Barn, which would help to relieve the overcrowding.

But another section of the community was dissatisfied with the way things were going on in the tramway service; thousands of drivers, maintenance men, blacksmiths, cleaners and the stablemen who fed and groomed the Company's eighteen hundred horses. These men worked for long hours at a low rate of pay, under shocking conditions. The D. U. T. Co., however, had a simple but effective way of dealing with anyone who showed signs of making trouble. He was sacked on the spot. In this way, a standard of discipline akin to martial law was maintained.

Tram drivers were the worst off. They started out at 6 a.m., often in wet clothes and remained at their posts for nine hours, exposed to hail, rain or snow. There was no such thing as a "dinner hour." A little boy or girl would meet the tram at its terminal "turn round," hand up a package of bread and butter and a billycan of hot "tay," to Daddy, and run away again, as if in fear of being caught. The driver had his meal on the driving seat while the tram was in motion, or during short stops.

One of the drivers, more daring than his fellows, obtained an interview with the Managing Director, Mr. Murphy, but, when he started talking about hardships and time off for meals, the only satisfaction he got was a piercing glance from the gimlet-eyed plutocrat and an infuriated "Dinner hours!!! What next?" And the interview ended.

After that, a man called James Connolly, a founder member of the Irish Social Republican Party, came into the news. He had been going around the tram depôts, trying to get the men to join his union but was met with little response. At this period, there was a nucleus of up to one hundred trades unions in Dublin, but few of them had enough prestige or membership to make any impression on the employers.

But there were many portents to indicate a change in the everyday life of the city. Reports about the advantages and growing use of electricity, especially when applied to public lighting and transport, were

Plate 1. U.S.S. Concord, on which Kyran Quaney served as seaman and diver in the United States Navy.
By courtesy of Smithsonian Institution and U.S. Navy.

Plate 2. U.S.S. Oregon, on which Kyran Quaney served as Electrician, First Class.
By courtesy of U.S. Navy.

Plate 3. Horse trams on O'Connell Bridge, 1896.

Plate 4. Cliften Robinson driving the first electric tram in Dublin on May 16th., 1896

Plate 5. Early Dalkey electric tram with tow-tram, 1897.

Plate 6. D.U.T.C. open-type four-wheeler in original colours at north side of Nelson Pillar, Sackville Street, Dublin, 1899.

constantly appearing in the newspapers. These were supplemented with statistics from municipal transport undertakings in London and other English cities, which proved beyond doubt that electric tramway systems had become popular and successful, wherever they were tried out.

At this critical stage in the history of Dublin's transport service, Cliften Robinson returned from his travels, to pick up the threads of a career that had been ignominiously cut short, when he was sacked by a Tramways inspector at Westland Row station, thirteen years before.

Cliften returned in triumph and established himself on a firmer footing. He had made a fortune in the United States and was well connected with a group of financiers and engineers, who had already taken part in the electrification of public transport systems in London, Birkenhead and Bristol.

Clifton had developed into a big business man in every sense of the term. Not only did he possess a dynamic personality and that all-important attribute of a company promoter, the "gift of the gab," but seemed to be endowed with the "gift of the grab" as well. From the time he arrived in Dublin things started to hum. He floated a company, obtained way-leaves to start an electric tramway service from Haddington Road to Dalkey and, with the pull he had among business men in the London Stock Exchange, succeeded in having Bills for rights-of-way rushed through the Houses of Parliament in record time.

Before the smug self-satisfied directors of the Dublin United Tramways Company had time to rub the sleep out of their eyes, Cliften Robinson, Managing Director of the Dublin and Southern District Tramways Company, had squads of men digging up roads, laying down tracks and setting up poles and trolley wires for a line of electrical tramcars, to run between Haddington Road and Dalkey.

A year later, on the morning of May 16th, 1896, Cliften Robinson, in the presence of the Lord Mayor and Directors of the Dublin and Southern District Tramways Co., drove the first electric tramcar out of the sheds at Ballsbridge and officially inaugurated the new passenger transport line from Haddington Road to Dalkey.

This line, afterwards known as the "Dalkey" line, was a great success from the start and the Company made an all-out effort to put as many trams as it could into service. The generating plant at Ballsbridge station was augmented to meet the increased demand for electric current, and, to make the best of the available power, the trams were to run in pairs during rush hours. The first car, worked by electric motors towed a trailer car without motors. The service was supplemented, until thirty pairs of trams were running on the system during peak hours.

The new Dublin and Southern District line became so popular, that between May 16th and August 1st, 1896, 400,000 passengers were carried between Haddington Road and Dalkey. There was a great shortage of tram drivers and Cliften Robinson, man of action that he was, trained men himself at night and in the early hours of the morning, when normal traffic was suspended.

But the old and reliable Dublin United Tramways Company had no intention of being run off the streets and out of business, by any new rapid transit concern. When the first shock of a competitor's encroachment, on territory it had marked out for future development passed, the Directors held an emergency meeting and decided that their best counter-move would be to open a new horse-tram line from Abbey Street to Haddington Road, to "feed" the electric line with passengers and incidentally, forestall any intention its rival might have, in the matter of extending the electric line towards the city and its potential threat to their interests.

The success of the new electric tramway caused gloomy forebodings in many Dublin business circles. Horse-transport formed the basis of many national industries, and experts prophesied that if it was abandoned in favour of modern mechanical traction, old-established businesses would be faced with extinction. The pros and cons of a possible change-over to electricity were debated in Chambers of Commerce and the general opinion was, that an all-Dublin electric public transport system would be good for the few and bad for the many. This opinion was shared by the Directors of the Dublin United Tramways Company and its associate business interests. A campaign was started, to prevent a possible extension of the new electric tram service, into the city proper.

The campaign took the form of writing letters to the newspapers, protesting against any further use of electricity, as it would cause untold unemployment, and sending delegations to the Dublin Corporation to demand that no way-leaves be granted to the new electric company, to extend its transport line towards the city.

As a master-move, the city Fire Brigade Chiefs were invited to express their opinions. They testified that it would be very difficult to manoeuvre fire escapes into position and extend them to high roofs or windows in streets where tramway poles, trolley wires and span wires would form an obstruction and that the risk of fire, spreading from one block of houses to another, would be greatly increased.

Within a short time, public opinion was aroused and everyone had something to say about the approaching transport crisis. Meetings were held by coachbuilders, blacksmiths, harness makers and others whose means of livlihood were likely to become affected. Businessmen, tramway officials, tradesmen and professional agitators joined in the chorus, calling on all citizens to make it clear to the Dublin Corporation and the British Houses of Parliament, that they were unanimously against any extension of electrical transport in what had become popularly known as the Battle of the Trams!

In the face of these protests, James Cliften Robinson, Managing Director of the Dublin and Southern District Tramways Company, introduced an application to the Dublin Corporation from the Bristol Electric Tramways Company, in which he was also interested, for "direct and untrammelled access to the centre of the city, to cater for the growing traffic from Dalkey, Kingstown, Blackrock and other south Dublin surburbs."

The die was cast. Cliften Robinson had been so successful with his previous applications for rights-of-way, that this latest petition in itself was enough to throw the Directors of the established horse-tram interests into a panic.

Thoroughly alarmed by the increasing publicity about the spread and popularity of electrified transport systems, the Dublin United Tramways Co., had prudently sent a delegation of shareholders to the United States, to find out the latest developments in electrical traction, and after visiting New York, Chicago, St. Louis and other cities, the committee reported that overhead trolley wire systems were in most general use.

The value of this report however, was somewhat discountenanced by the findings of a committee of American experts:

B. S. Flanders, City Electrician, Boston.

Wm. Brophy, Expert for the Commissioner Wires, Boston.

John B. Barrett, City Electrician of Chicago.

Morris B. Mead, Superintendent of the Bureau of Electricity, Pittsburgh.

All were agreed that:

"While the overhead single or double trolley system may be permitted on railroad lines running through rural districts and possibly, in the sparsely populated towns and cities, no overhead system should be tolerated in the mercantile, manufacturing or densely populated sections of the cities."

These conflicting reports were disturbing, but then — a crisis had arisen in the development of public transport systems, and something had to be done about it without delay.

William Martin Murphy and his colleagues in the D. U. T. Co., now realised that they could no longer ignore or impede the modernising influences which were causing changes everywhere and that they were face to face with a competitor who could, by extending a better transport system into their hard-won territory, gain the goodwill of the public and run them out of business.

It was at this critical stage in the affairs of the Company, that Martin Murphy displayed those characteristics of leadership that were to endear him for ever in the minds of Dubliners who worshipped at the shrine of private property and success, and make him detested by the proletariat, who regarded him as a ruthless autocrat "out to crush the working man and keep him down."

He first arranged a meeting with his rival, Cliften Robinson, who had to admit that the Dublin United Tramways Company would have first claim to provide electrified transport on all routes already served by its horse-trams, and in consequence the Dublin and Southern District Company would never have a chance to extend its thirteen miles of track citywards.

It had been proven time and again, that short-haul passenger transport lines never paid in the long run, and when all the cards were on the table, Cliften was forced to acknowledge that his Company would in the course of time, become bottled up and uneconomical.

When the meeting concluded it was announced that the Dublin United Tramways Company had acquired the whole share capital of the Dublin and Southern District Company in a merger that became known as the Dublin United Tramways Company 1896 Ltd.

After selling out his interests in Dublin, Cliften Robinson returned to England and became Managing Director of the Bristol Electric Tramways Company. He was knighted for his services in public transport by Queen Victoria, a few months before she died.

William M. Murphy was now looked upon as the leading business advisor in the city, and was called on to rule the destinies of half a dozen companies. The directors and shareholders of the D. U. T. Co., had the greatest faith in his judgment, and when it came to making decisions on matters of planning or policy, his opinion counted above all others. He was very much interested in electricity and could foresee a wonderful future for it in Dublin.

At the 31st Ordinary General Meeting of the D. U. T. Co., Dr. Carte, the Chairman, announced that the residents of the Clontarf area, had unanimously signed a manifesto approving of an electrically operated line between Annesley Bridge and Clontarf, a distance of approximately four miles.

This was the cue for William M. Murphy to outline his master plan for the electrification of the whole of the Company's horse tram network and he proposed that a start be made on the Clontarf line, which passed through a prosperous area that had great potentialities for future development. His proposal was unanimously accepted.

Before the meeting ended, plans for the construction of an electrical generating station in Clontarf were approved and the recruitment of a staff of engineers authorised.

It was around this time that Kerr returned from Limerick to Dublin, and rented a new house on the Richmond Road, not far from Annesley Bridge where the new line was to be constructed.

The introduction to Dublin of electricity in the year 1896, caused a general upheavel in the lives and habits of its citizens. Two great industrial changes were impending. The Dublin United Tramways Company was about to switch over from horse-drawn trams to electrified rapid transit, and the Dublin Corporation had made plans to provide electricity for general street lighting and private houses.

Kerr Quaney, fresh from his successful assignment in Limerick, felt that his prospects of an electrical career in his own country were good, and he didn't lose time getting in touch with the principals of the proposed enterprises. He first obtained an interview with Mr. J. Toule, a young and brilliant engineer who was Martin Murphy's right hand man and who, like himself, had learned his business in America. They were talking only for a few minutes when his interviewer interrupted him:

"Look here, Kerr, you're the man we want. Y'know this Company'll have to float on credit for a few years before the money starts coming in. I'd like to give a man of your experience and capabilities a big salary, but all I can do is to offer you something small for a start. If you accept and come in with us — why y'can't lose and the salary will grow up with the importance of your job."

Kerr hadn't received a reply to his application for a post in the Dublin Corporation and was feeling worried. He couldn't afford to remain out of work for long and the prospect of a long-term post was tempting. Acting on impulse, he accepted Mr. Toule's offer.

"There's a power station being built at Clontarf. Take charge of it," said Mr. Toule. "The generators are due in about a fortnight. Meanwhile, get all the labour you want and appoint a few gangers. You'll have to see to lines, tracks, poles and everything. The best of luck."

With this simple briefing, Kerr started on his career as engineer in the Tramway company. Before the week was out, he received a letter from the Dublin Corporation offering him a permanent and pensionable post, at nearly twice the money. He looked at it sadly and threw it into the fire. He had given his word and wouldn't go back on it.

"You did the right thing, Kerr," said his young wife.

It was with a sense of disappointment and frustration that Kerr made a start on the supervision of the constructural work at Clontarf. His own knowledge and experience, backed by the opinions of several American experts, told him that overhead trolley wires were a constant source of danger to city crowds. Mr. Toule himself had said he would have preferred a central "live rail" system that would be far safer, but he had no say in the matter.

The decision to use overhead wires had been made by the Board of Directors, on the advice of a delegation of the most affluent and influential shareholders which had returned from the U. S. A. after a sight-seeing tour at the Company's expense. On the strength of it, the Company had made irrevocable plans and commitments.

Mr. Toule maintained that the shareholders who were responsible for the decision, knew as little about electricity as a baby would about metaphysics — but there it was — they had the "say" in the matter.

Overcoming his forebodings and misgivings, Kerr threw himself into his task with the energy of a young man who was out to make a name for himself in a highly competitive community, where brains and skill were the passports to lucrative executive posts.

He had four miles to travel from his home on the Richmond Road to Clontarf power station, so he bought a bicycle and pedalled to and from work each day.

Labour was cheap and plentiful and only the best men were employed. The pay was fourteen shillings for a fifty-four hour week. The construction of the power station was soon under way. A boiler house with a fifty-foot high chimney stack was built to provide power for the steam-driven electric generator.

The three-mile stretch of coast line, from Fairview to Clontarf was a scene of feverish activity. Men worked around the clock sinking thirty-foot tubular steel poles, six feet deep, into the ground in pairs, one on each side of the road, fifty feet apart. Wiremen followed up, stretching span wires fitted with double porcelain insulators to prevent leakage of current to earth.

Trolley wires were suspended from the span wires by means of long copper "ears" which were grooved on the underside so that the top half of the wire fitted into the groove, and both were then soldered together.

The whole sea front presented a satanic picture at night: Tar boilers, billowing smoke, were stationed every hundred yards or so, and between them gangs of men dug up the road — ready for laying tracks. Blacksmiths worked at coke braziers, heating and tempering crowbars, pickaxes and chisels blunted from dressing the hard stone setts. Smoke-blackened men, walked around in the light of paraffin flares with buckets, pouring boiling tar between the joints of the setts, bonding them together to hold the tracks securely in the road.

The generating plant in Clontarf was operated by a stationary steam engine. A dynamo fed electricity to the overhead wires, which was collected by the trolley wheel and ran down through the tram by conductors to the motors. When the current passed through and activated the motors, it ran to earth via the tram tracks. If the overhead wire did happen to break, it would be noticed at once, by the engineer on watch in the power station and he would cut off the current by manipulating a hand-operated isolator switch.

During the construction of the Clontarf line, Kerr had little time for home life or social engagements. Workmen were paid by the hour, but the engineers had no fixed hours and were expected to be on duty on Sundays, Bank Holidays and even on Christmas Day. The word "overtime" was taboo among senior officials of the Company. A week's holiday in the year was supposed to make up for all the extra hours of duty and it was a point of honour that executives wouldn't consider taking holidays, if they were likely to interfere with the Company's interests.

Life at the time was one continuous round of work and increasing responsibilities. There was another addition to the family. The young couple were happy enough in their struggle for a living. With five golden sovereigns coming in every week, Kerr was regarded as a class of millionaire by some of his neighbours and his wife was able to afford a maid. However, the cost of living was rising steadily and the money seemed to go as quickly as it came in.

Work continued on the Clontarf line day and night throughout the year 1896, and early in March, 1897, arrangements were being made to put the line into service. The generating plant in the power house had been tested and everything was in running order.

Kerr Quaney was responsible for the safe working of the whole line, and only one thing really worried him. He had personally trained the men who put up the overhead wires and had a hunch that some of the work might be faulty. There were no automatic cut-outs in those pioneering days, and if an overhead wire broke, it fell to earth and burned itself out in a tremendous spluttering flame until the wire melted clear of the ground.

To satisfy himself, he arranged for an inspection trip on the top of a tramcar, to examine the wires before passing the line for service. The early trams had seats on top and the passengers were exposed to the weather. The inspection trip started at dawn on a chilly March morning. Both driver and conductor were on their first run; two other men were there to help.

It was slow going, starting and stopping every fifty yards or so for Kerr to check the span wire insulators and the soldered joints between the trolley wire and the "ears" that suspended it; but it was good practice for the tram crew. Everything seemed perfect until the tram got to the top of old-fashioned, hump-backed Annesley Bridge over the canal, when Kerr thought he could see a fault in the joint between the wire and the "ear." He stood on top of the back seat and gripped the wire with both hands to make a closer inspection. He was at an awkward angle and to get a better view shouted to the driver "go ahead two feet."

The driver released the powerful hand brake and immediately the tram moved quickly down the slope of the bridge, leaving Kerr dangling from the wire about twenty-five feet above the cold stone setts. The driver panicked and couldn't stop the tram. When at last he did succeed, he couldn't back the tram to collect Kerr, because the current was switched off at the "feeder box."

Kerr could never tell how long he was hanging from the wire — it seemed hours. He was on the point of letting go when he saw the four men stretching out an overcoat for him to drop into. They caught him so well that he didn't get a scratch and was able to carry out the rest of his inspection as if nothing had happened. On March 20th, 1896, the electrified line from Clontarf to Nelson's Pillar was opened and the system operated faultlessly from the start. The permitted speed of trams was eight miles an hour.

1897 was a year of intensive preparations for the electrification of existing horse tram routes, and the nature of the work caused a lot of disorganisation and obstruction to owners of business premises. Gangs of men seemed to be endlessly digging up the streets, laying tracks and excavating holes for poles and when they finished other gangs came along digging up the footpaths to lay underground cables and erect massive cast iron "feeder boxes" on every half-mile of the route. The horse trams plodded patiently through the chaos, maintaining passenger services.

Only the promoters who hoped to profit by the change seemed to have any enthusiasm for the electrification scheme. As the work progressed and the traffic jams became greater, public animosity spread, until the name of William Martin Murphy and his electrical tramway scheme was soundly cursed in every street where the pole squads and tracklayers made their appearance. But the D. U. T. Co., had put every penny of capital into the venture and had to go through with it or lose everything. So the work went on doggedly from street to street in the face of unprecedented disapproval and resentment.

Following his successful establishment of the Clontarf line, Kerr was appointed Engineer-in-charge of the power station. This was an honorary promotion as it carried no increase of salary but his name as an engineer was made with the Company from that time on.

The Company installed its own telephone system as the electrical network spread. Each "feeder box" had a telephone instrument and the engineers were supplied with heavy bolt-type master keys so they could phone from one part of the system to another in the event of a breakdown or accident.

Without looking for further advancement, Kerr automatically became second-in-charge of the whole system. His senior, Mr. Toule, was over in London most of the time in connection with the plans of the proposed Central power station, and there was no one else to deal with complicated technical queries that arose when he was away. When any vexing problem cropped up, those in charge usually resorted to "get in touch with Quaney." So he was continually in demand, at points where the work was stopped for lack of technical skill and knowledge.

18

EARLY IN March, 1898, Mr. Toule, accompanied by H. F. Parshall, an eminent British engineer and designer of the new Central power station for the Dublin United Tramways Company at Ringsend, returned from London with the plans ready for acceptance.

The vast engineering scheme was an outsize proposition for a private firm, probably the greatest ever attempted in the history of private enterprise in Europe. The choice of site in itself was a brilliant piece of planning on the part of an engineer who was big enough professionally to visualise the whole project in detail and practical enough to see it through and achieve the great ambition of his career.

The foundation of the main building was set in line with the south side of Ringsend Dock, a large tidal basin fed by the River Liffey. Ships of up to a thousand tons burden could enter the basin to load or discharge cargo, and the dockside was well equipped with coal grabs and mobile cranes capable of dealing with massive loads. The site was suitable in every way for the delivery and handling of cargoes of structural steel, heavy machinery, poles and tram tracks — not to mention the 50,000 tons of coals a year it was estimated the furnaces would burn when the tramway system developed to its peak efficiency.

The main power station was 808 feet in length by 200 feet wide by 80 feet high. The roof was of structural steel members and trusses, with a single span of 200 feet. Light was admitted through panels of yellow roughcast glass in the roof and walls. All the structural steel was manufactured by Reter and Conoly of Pittsburgh, Pennysylvania; assembled in their factory, taken to pieces again and shipped to Dublin for re-erection. The sections were dispatched in order of assembly so that no time would be lost sorting out, and once started, the building took shape as if it was being built in America.

The site was several acres in extent. All the buildings were set out and excavation for foundations began before the month of March was out. Two hundred and fifty men were employed continuously from the time the work started. In June of the same year the construction of the power house was well advanced, and sections of the specially designed coal tower began to arrive. This was a complete distributing system in itself. The coal was taken from the ship by mechanical grabs, hoisted to the top of the coal tower and dumped into waiting bogies. The bogies rolled along elevated runways and automatically tipped their contents into twelve vertical chutes to feed the furnaces under the boilers, eighty feet below. All the firemen had to do, when the furnaces required refuelling, was to pull a chain and the coal tumbled down into a hopper over the furnace door. The fire boxes had endless heavy steel chain grates that kept moving around the clock to prevent the formation of clinker.

At the half-yearly meeting of the Dublin United Tramways Company on July 27th, 1898, the Chairman, Dr. Carte, J. P., announced that the construction of the central power station at Ringsend was proceeding to schedule and that when finished, it would be the finest electrical power station in Europe. He then went on to say that the laying of tracks, erection of poles, trolley wires and underground cables was going ahead on the north Quays, and that the Donnybrook, Rathmines and Sandymount lines had been electrified from Ballsbridge power station.

Reports about public dissatisfaction on routes where horse and electric tramcars met and passengers were asked to change from one type of vehicle to another were read out. The mix-up with tickets at these points caused another rumpus, so it was decided to issue tickets that would be available on either horse or electric trams until the overlapping routes were reorganised.

During the winter months of 1898 half of the main traffic routes in Dublin were in the throes of reconstruction. Existing horse tram tracks were being taken up and relaid to suit the new gauge for electric bogies. Poles and trolley wires, nicknamed "sky-wires" went up in street after street.

The proprietors of small businesses complained about disruption and loss of trade. Long and abusive letters appeared in the daily papers and public indignation mounted as the chaos increased, but the work went on inexorably day and night. The "sky-wires" went up in all directions as the easy-going era of stage coach transport was ousted by electricity.

In the spring of 1899, the walls and roof of Ringsend Power Station were up, and the wonderful red-tiled floor was laid. Six electric generators, made by Allis and Company of Milwaukee, U. S. A., were installed in their massive cast-iron casings. Each engine generated seven hundred and fifty horse power and the fly-wheels — nineteen feet in diameter by a foot thick — weighed thirty tons apiece. Twelve Babcock and Wilson boilers arrived from Manchester about the same time. The constructional and assembly work at Ringsend was now at its peak. At this stage H. & G. Martin, building contractors of Dublin and Belfast, started work on the forty-foot square solid concrete bases for the two giant chimney stacks, popularly known as the "Twin Sisters."

The stacks were constructed from shaped steel plates with interior firebrick linings ten feet in diameter. Each chimney was two hundred and twenty feet high and both were erected in the short space of seven weeks, an amazing feat of craftsmanship and efficiency.

It was said that the man who lit the first fire opened a steel door in one of the bases to test the draught and see how the fire was going on, and the suction of air was so great that it nearly carried him up the chimney with the smoke.

A great network of overhead wires and underground cables now converged at Ringsend. One morning the Consulting Engineer, Mr. Toule, rang up Kerr — on duty in Clontarf.

"We're nearly ready to start connecting up the main switchboard, Kerr. Could you manage to leave Clontarf for a while and do something here in Ringsend?" There were two young watch engineers trained to take over and everything looked satisfactory for the time being.

"O.K., Mr. Toule," he replied, "guess I'll be along shortly."

On arrival at Ringsend he was heartily greeted by his chief.

"I'd like you to take over here for a while, Kerr, there's no one else I know who could tackle it," he said. As they stood there in conversation and glanced around the power station, everything seemed to be on an immense scale, even to Kerr.

On a ten-foot high steel platform, fifty feet long, a black vulcanite panel nearly seven feet high ran the length of the platform, ready for mounting the main

Plate 7. Electric trams at G.P.O. and Nelson Pillar, Dublin, 1900.
(National Library of Ireland).

Plate 8. Modern D.U.T.Co. trams at Donnybrook, Dublin.

Plate 9. C.I.E. bogie car.

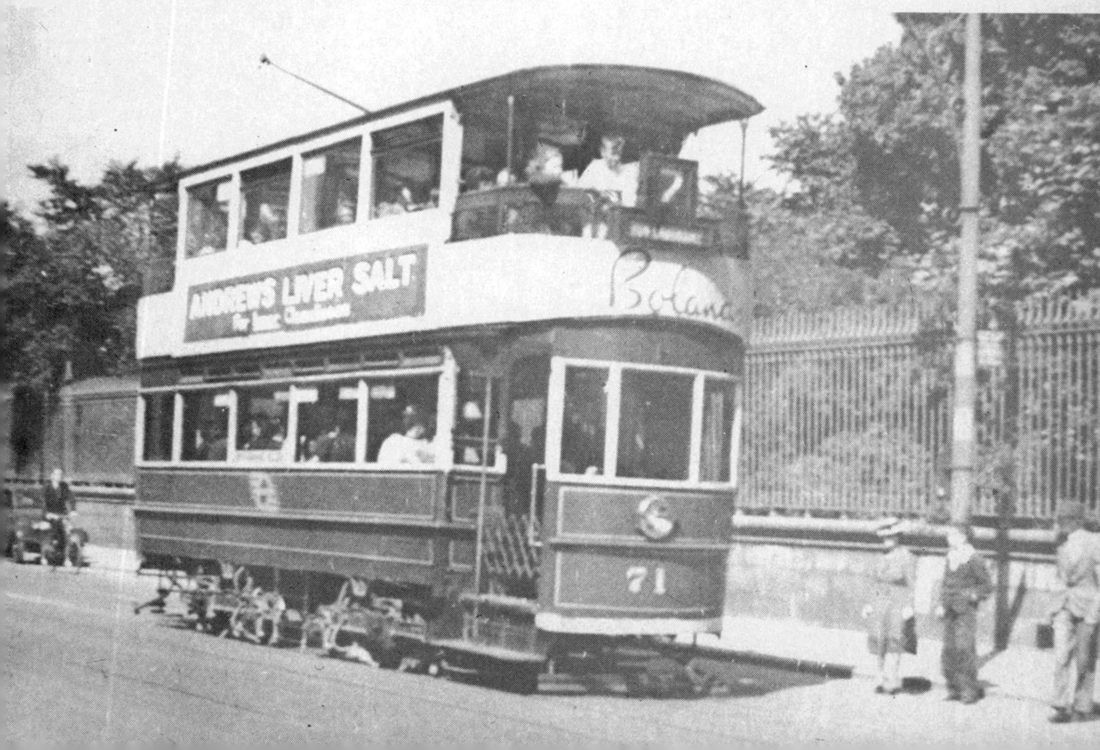

Plate 10. Dublin and Blessington steam tram. The last steam tram ran on 31.12.32.

Plate 11. G.N.R. Hill of Howth tram. The last tram ran from Sutton on 31.10.59.

Plate 12. D.U.T.Co. standard four-wheeler, 1940.

Plate 13. Model of C.I.E. luxury tram No. 280, by Mr. Cyril L. Fry. One of this type was the last tram to run from Nelson's Pillar on July 9th, 1949.

switch gear and fuse boxes. Cables for a dozen tram routes lay around on the floor snaked in coils and bights, intermixed and interwoven as undergrowth in a jungle. The job was to get them sorted out and connected to control switches so that any of the lines could be made "alive" or "dead" at the pull of a switch.

"It'll take at least three months," replied Kerr, after sizing things up.

"Righto, go ahead, I'll leave it to yourself," returned his chief.

Those were halcyon days for the young engineer, full of ambition and completely absorbed in his work. The main thing was to get the system on wheels and see that it paid its way. Already he was planning ahead, visualising problems that were bound to arise when the Company would operate a tramways service second to none in Europe, or America. Spurred on by these thoughts, he had the Company's interests at heart as much as if he owned it, though in reality he was too poor to invest in a single share.

Kerr was a tireless inventor, ever on the lookout for an opening to introduce an industrial improvement or machine, that would work more efficiently than the one that was in use, and he felt sure that in the vast industrial undertaking in which he was employed, an opportunity would arise in which he could exploit his flair for invention.

When electric trams started running in Dublin, a wave of accidents occurred that turned a large section of the public against them. There were no regular starting or stopping places and when a crowd of passengers formed in a street, the driver stopped to pick them up. Sometimes a passenger wanted to get off a tram in a hurry and the conductor wasn't within speaking distance. People knew how to stop a train by pulling the communication cord and thought the same would apply to a tram.

This misinterpretation of procedure caused an accident on the Clontarf/Nelson's Pillar line on the first of August, 1899. A passenger wanted to get off the tram at Amiens Street station and the conductor wasn't within hailing distance, so he pulled down on the trolley rope and let go. The trolley sprang straight up into the air. Carried along at speed by the momentum of the tram, it crashed into the first span wire it met, carried it away and brought down the trolley wires on top of the tram and passengers. The tram instantly became "alive," sparks flew in all directions and three passengers were taken to hospital suffering from shock and burns.

There was pandemonium in the neighbourhood that almost ended in a riot when the news spread. The police took sides with the incensed passengers and arrested the conductor for not being on his platform when the passenger wanted to get off. The Clontarf line was blocked for a whole day and the Company incurred a serious loss.

This accident required a lot of explaining away and there was much talk about bringing in safety regulations but the outcry died a natural death when the Company said it would consider employing two conductors, one to look after the platform and trolley, and the other to collect the fares. This was only a diplomatic gesture and things remained as they were until the incident was forgotten.

However, as a result of this mishap Kerr Quaney suggested the introduction of fixed stopping places two hundred yards apart, to be indicated by painting the upper section of every second pole white. This proposal was immediately put into practice and the number of getting on and off accidents was considerably reduced. A big economy in the amount of electricity used was also effected.

In December, 1899, the central power station at Ringsend was brought into service. The opening of this station made Clontarf generating station redundant but the power stations at Blackrock and Ballsbridge were still in use.

At the seventh Ordinary General Meeting of the D. U. T. Co., in February 1900, William Martin Murphy, the Chairman, announced that the electrified lines did not account for more than half of the gross receipts from passengers.

On July 31st, 1900, at the eighth Ordinary General Meeting of the D. U. T. Co., Mr. Murphy confirmed that there were only twenty-two horses left out of between one thousand five hundred and one thousand six hundred, and that these horses were working on the Haddington Road/Sandymount line. Several electrified tram routes, including the Donnybrook /Phoenix Park, Glasnevin/ Dolphin's Barn and College Green/Inchicore lines were opened up the same year. The Company's overall plan to operate three hundred cars on one hundred miles of street track was about half-accomplished.

Crowds of intending passengers found it difficult to identify the routes of the new fast-moving trams, especially at night-time around the city centre. To simplify this problem, Kerr suggested the introduction of coloured lights and shaped illuminated glass signs at the motorman's end of the tram, so that passengers could easily identify their homeward bound trams from a distance. This idea was put into use, and remained a feature of the Dublin trams during all their years of service.

The growing use of electric light and tramcars was a topic of conversation all over the country, especially among those who were in favour of "advancing with the times." The Limerick Chronicle of November, 1898, reported:

A professor addressing the Literary class of the Protestant Young Men's Association said;

"Yes — the time may come when amid the brilliant illumination of the electric light you may see a modern electric tram stealing noiselessly down the main street of the historic city of Limerick."

The early electric trams in Dublin were anything but noiseless. They swayed and screeched at bends and often jumped the tracks. They were built for utility and hard wear, and the standard of comfort was austere. Seats made from hardwood slats stretched the full length of the passenger compartment. Passengers sat in a row on each side of the car facing one another. A row of carbon filament lamps on each side of the tram threw a sickly yellow light on the passengers. The lights glared and dimmed according to the amount of electric current used by the motorman driving the car.

Up on top there were two rows of seats, each seat to accommodate two passengers. There was a single seat on each side of the trolley standard. The backrests could be moved back or forward so that the passengers could face the breeze or have it at their backs.

George William Russell — "AE," the poet, described the trams as "Galleons of the streets." On a wet stormy night when the theatre crowds were making for home, they were a truly stirring sight converging at Nelson's Pillar. Up to a hundred trams would be on the move in Sackville Street, bells clanging, crossing and recrossing tracks, each in its halo of yellow light with sparks flashing from the trolley head, distinctive blue, green, yellow or red identification lights and illuminated multi-coloured signs claiming special attention. When heavily loaded the trams dipped and swayed like trawlers on a choppy sea and the emergency fenders under the conductor's and motormen's platforms (to push anyone who fell on the tracks clear of the wheels) made a mighty clatter as they scraped and crashed on the stone setts at every dip of the tram.

Once Ringsend power station was commissioned and in regular working order, the strain of overwork eased, and the engineering staff had time for a breather. The

Company was paying 6% on Ordinary and the same on Preference shares and there was every sign of prosperous times ahead for the firm and everyone connected with it.

One by one the old-fashioned, easy-going trams faded from the streets and on the 13th January, 1901, the last horse tram left Haddington Road for retirement to the sheds at Sandymount. A great crowd had gathered to give it a hilarious send-off. In the excitement that followed, a man fell on the track and the wheels of the car passed over him causing fatal injuries. Thus the joviality was suddenly overshadowed by tragedy as the curtain fell on a system of transport that served the citizens of Dublin truly and well for over thirty years.

At this stage, the engineering staff was reorganised and Kerr Quaney was appointed chief overhead and cable engineer for the whole system. The new post carried a small rise in salary and a big increase in hours of duty and responsibility. A horse-drawn, high seated inspection trap (a fine turnout with a uniformed driver) available at any hour of the day or night, went with the post.

Although the new position added considerably to his prestige and importance, Kerr would have preferred any other engineering appointment, but home commitments weighed heavily on his mind and he had no wish to jeopardise his future by asking for a change.

He had been opposed to overhead trolley wires from the start, and to be suddenly made responsible for a whole cityful of them imposed a mental strain that he felt he could hardly stand up to.

Three anxious years passed on a never-ending inspection of the overhead wires. They broke down frequently from various causes, sometimes from extra friction over heavy duty routes, in other cases they were damaged by badly grooved trolley wheels and again the wires became tired and brittle from swaying in the wind. After years of patient study, one fact emerged clearly — most break-downs occurred at one point and that was where the wire was attached to the "ear."

Section insulators had been invented and were installed in all "feeder boxes" so that when an overhead wire broke and fell to the ground, a circuit breaker automatically cut off the current and that section of the line became "dead." It remained so until the wire was repaired and the circuit breaker reset.

But the crucial danger was, that when an overhead wire broke, it could fall on the passengers on top of the tram or on a crowd in a street, and kill or injure several people before it touched ground and was "deadened," by a circuit breaker.

Although there had been numerous break-downs and miraculous escapes, there were no fatal accidents as yet, but as the reports of "near misses" accumulated accident insurance rates went up, until the D. U. T. Co., had to foot an insurance bill that kept rising annually as the claims for accidents increased. At one General Meeting the Chairman had to announce that the Company seemed to be considered fair game for claims of any sort.

Accident risks increased in proportion to the city crowds, and whenever a Royal visit, political meeting, big funeral or parade was imminent, the Overhead and Cable Engineer had a run of sleepless nights until it was all over.

Then one day Kerr had a flash of inspiration, a brilliant idea that would, if patentable resolve all his worries and put him on easy street for the rest of his life. It was a safety device that would hold a broken trolley wire in position and prevent it falling to the ground and causing damage. He kept thinking it over for days, astounded that such a difficult problem could be solved so easily. Now, sure beyond all doubt that his invention would be successful, he mentally christened it the "Anchoring device."

But it was one thing to have a practical idea and another to get out a patent to protect it from being copied. It was an era of invention, especially of electrical devices. Getting out patent rights would entail a lot of research, making drawings of the design and filing an application for provisional protection all of which would cost money.

Kerr hadn't a penny to spare. The patent laws required that he wouldn't divulge particulars of his invention to anyone other than a partner and in the event of a partnership each would have equal shares in any profits accruing from the sale of the device.

He believed that the value of his invention was high because it was original and would prevent loss of life. If patented and exploited on an international scale the device could be worth millions. He was naturally in a fever of haste to get the idea protected. Being first with an application for a patent sometimes meant everything, as it was of common occurence that several people hit on the same idea at the same time.

He decided that his best policy would be to find a partner who would undertake the risk of paying all expenses in return for half a share of the profits. There seemed to be no other way out.

He went to see the Rolling Stock Engineer, a friend he could trust. Mr. Kirwan was so impressed with the invention, that he agreed to become a partner on the spot. Within a week the drawings were ready and an application for a provisional patent was filed under the registered name of the "Kirwan/Quaney Anchoring Device."

The directors of the Company, acutely aware of the rising cost of accident insurance and their responsibility for the safety of the public, as the tramway lines extended out to the suburbs, were greatly relieved to hear that one of their own engineers had invented a device that would dissipate much of their anxiety and possibly immunise the Company against big accident claims. They wholeheartedly agreed that the "anchoring device" should be given a full working test at the Company's expense and that the Chief Engineer should make a report on its performance.

Within a few months the invention proved itself an outstanding success, and became known throughout the city of Dublin as the K. Q. patent. At the next Ordinary General Meeting the directors were so enthusiastic that they wanted to have K. Q's fitted on every line in the system, but this was ruled out on account of the cost. They finally agreed to begin to use the apparatus by degrees — on all new lines as the wires were put up, and on existing lines when they broke down from wear and tear. All were of the opinion that the invention was wonderful and that when every trolley wire in the system was "anchored," the danger of falling wires would be gone for ever, and the safety of the public assured. The overwhelming success of his invention was a great source of pride and gratification to the young engineer and his family. Driving around the streets daily in his smart inspection trap, he became one of the best known men in Dublin. Thousands knew him by name and people used to say "a man on the upgrade like him will surely become a millionaire."

With the prospects never looking so good, the family hoped and prayed for better times, and Kerr waited patiently for news about an award for his invention but weeks passed into months and no word came. Disappointed about the way things were turning out, he went to see the Secretary, in the Company's head office in Sackville Street.

The Secretary, middle-aged and rotund sat at a long table in the Board room, exuding prosperity, urbabity and benevolence. His manner indicated that the visit was not unexpected.

"Well, Mr. Quaney, delighted to see you, I've been waiting for you to call. The Board and I heartily congratulate you on the success of your invention and we are all proud to be associated with you in our service to the Company. Of course it's hardly necessary for me to remind you that the Company is still struggling to get on its feet and that owing to widespread commitments it has yet insufficient funds put by to show practical appreciation to individual members of its staff — even for such outstanding services as yours. However, Mr. Quaney, you may rest assured that your exceptional contribution to the establishment and development of the Company will never be overlooked."

The pause that followed implied that the interview had ended. Kerr walked down the steps into Sackville Street in a state of shock. He had at least expected a royalty on each of the K.Q.'s used, or a substantial bonus, but to be put off with a vague assurance that his services would never be overlooked, was more than he could stand. He felt that he was being done out of a just award and for a moment toyed with the idea of resorting to law, to seek retribution.

However, after a walk he dropped the idea. Instead, he went to have a chat with his partner in the patent, who was equally upset when he heard about the result of the interview with the Secretary. As yet, only a Provisional Patent had been taken out. An application for Full Patent rights would require a lot more research and money, meanwhile the chances of someone else hitting on a similar idea and patenting it multiplied daily as electrical tramway systems were brought into use all over the world.

As they talked things over, both felt that as founder engineers of the electrified tramways system, their future careers were tied up with the interests of the Company and that their best policy would be to let the firm have the benefit of the patent until a reserve fund was built up and then apply for an award at an appropriate time.

19

PEOPLE LIVING in Dublin between 1900 and 1906 saw a great change take place in the lives and habits of its citizens. Smart carriages and musty four-wheeled cabs were still prominent in the flow of traffic. Sprightly jaunting cars, that catered for racegoers and week-end trippers, stood waiting for hire at stands in the centre of the city.

Stout waggons, furniture vans, milk-carts and parcels delivery vans, formed a large quota of the horse-drawn traffic, but landaus, broughams and other imposing equipages were definitely on the way out. It was no longer fashionable to ride on horseback.

Electrified transport was partly responsible for the change. Although the speed of tramcars was not supposed to exceed eight miles an hour, the law was ignored by drivers and inspectors of the D. U. T. Co., aided and abetted by the police. No tram driver ever appeared in Court for exceeding the speed limit. The modern craze for speed had started, and the tempo of city life had stepped up accordingly. All industry benefited from the introduction of a new and efficient transport system that enabled workers to get to and from their places of employment punctually.

The Dublin Corporation had opened up a huge generating station on the Pigeon House Road and was supplying electric light to factories, shops, offices and private houses. The main streets were illuminated brilliantly if not economically, by sizzling arc lamps after dark. The whole city was developing rapidly and the standard of efficiency in industry rose daily. Railways and canals had opened up provincial trade routes throughout the country and big business boomed as it never had before.

During that period, the citizens of Dublin, with the exception of a small group of political cranks, were Anglo-Saxon in custom and outlook and almost reconciled to British rule. Although the Church of Ireland was the government-sponsored religion, and most of its adherents were well organised, well-to-do Protestants, the overwhelming majority of the population was Roman Catholic and only a very small number of outsiders belonged to other creeds and beliefs.

The Protestant minority was stable and influential, the Roman Catholic masses poor— desperately poor. There was little bigotry. Religious opinions were respected and tolerated by both sides, but if a question of employment or helping a person was being considered, the applicant's religion would in all probability be a point in favour — or otherwise.

In the course of his duty as overhead and cable engineer, Kerr Quaney learned a lot about old Dublin. His driver Michael, who had once served in the British Army, and used to refer to him in conversation as the "Colonel," knew every street, alley and short cut in the city. When they were out in the high-backed trap, clattering through the dark deserted streets on night duty, Michael was good company.

"That's the house Lord Edward Fitzgerald was killed in," he would remark as they passed a gloomy old building near the City Hall, or further on:

"Robert Emmett was hanged there. Look at the marks where the scaffold was" — pointing his whip towards St. Catherine's Church in Thomas Street; or perhaps:

"Napper Tandy, the United Irishman, used to be around here in his day," as they took a short cut through the Coombe. He would continue to repeat snippets of history, until Kerr began to fancy that the ghosts of old Dublin walked the streets again, or peered furtively from shadows cast by the flickering gas lamps.

The more he saw of the Dublin slums the more he likened them to the housing conditions of the workers in New York. The cause was obviously the same. Industry was unable to absorb the available labour and mass unemployment increased as financiers consolidated their gains and created more redundancy by introducing labour-saving machines everywhere they could. Dublin seemed to be in the grip of a social malady for which there was no cure except that recommended by socialists and anarchists. The deeply religious masses of the Dublin workers wouldn't have anything to do with these foreign idealogies as they clashed with the basic teachings of their Faith.

Of the 300,000 people in Dublin at the time, over half lived in dark, oppressive slums. Most of the plague spots were within a mile of Nelson's Pillar. Middle Gardiner Street, Church Street, Summerhill and the Coombe were notoriously squalid, and experienced social workers rated them worse than any of the slums in London or Glasgow.

The cold, damp Irish climate added to the sufferings and misery of the poor, tuberculosis, rickets, typhoid and insanity and other diseases caused by overcrowding were rampant, and took an ever-increasing toll of life annually, yet despite the fact that the mortality rate was the highest in Europe, the wretched people multiplied like flies, and produced another generation, that fought for survival with the tenacity of microbes.

The average working man had a family of ten children and families of twelve or fifteen were quite common. Most working class families lived in a single tenement room where the never-ending functions of rising, dressing, cooking, eating and going to bed again were carried out under the most trying conditions of overcrowding. Yet in the midst of such chaos and discomfort, a family would not hesitate to take in and give temporary shelter to another family that suddenly lost a breadwinner, or was evicted for non-payment of rent.

The poor shared their trials and sorrows with true Christian humility, and the love of God was manifested by a tiny altar at which a perpetual lamp burned in the corner of the most squalid room. Religious organisations dispensed charity regularly and the Poor Law Guardians issued tickets for Outdoor Relief. This came in the form of money, coals, clothing, rations of food and other necessities that helped to prolong life and maintain hope.

A large proportion of the working class income was squandered on drink and tobacco. Porter cost twopence a pint, whiskey or rum fourpence a glass and cigarettes were five a penny. A labourer was paid fourteen shillings a week, with sixpence an hour overtime, tradesmen could earn a pound a week. As is usually the case when a hundred people are looking for every job that's going, there was no security regarding tenure of employment. The general rule in employment was that if a man looked as if he didn't like it, he got the sack.

Despite the widespread and degrading poverty, a high moral standard prevailed and ninety-five per cent. of the slum dwellers could claim that they lived in "respectable houses." Organised vice and crime was practically non-existent and the Dublin Metropolitan Police had a high reputation for fair dealing and moral integrity.

In 1906 Edison's new Living pictures were being shown at the "Rotunda" at fourpence a seat, but the terrible and more realistic drama of life in the slums could be seen any Friday or Saturday night for nothing. These were the two days in the week on which Dublin's poor had enough money to settle up with the pawnshops and engage on a limited spending spree.

When the parents were out enjoying a brief respite from the awful realities of their existence, the swarming children of the slums, out of the range of scolding tongues and rough handling for the time being, set out to do likewise. They rose to the occasion in any sort of weather and played hop-scotch on the pavements, sang "ring a ring o' roses" or hitched skipping ropes to the lamp standards, slipped into a noose and swung around in circles, until a waiting queue of mudlarks pulled them out and insisted on having their turn at the diversion. On gloomy wintry evenings the soot-laden air of the streets was little better than the foetid atmosphere of their tenement homes.

Drink brought the slum dwellers down to an even lower degree of misery than when they were sober, and at 11 p m. when the pubs spewed out their wretched, maudlin crowds of men, women and babies-in-arms, the evening's carousal invariably ended up in a series of family recriminations and brawls. Attracted by the noise, little children, waiting for daddy and mammy to come home with something to eat, stood around in their bare feet scantily clad, crying in vain before dispersing to their vermin-ridden beds.

On Monday mornings the trek to the pawn shops started again and with the loan of a couple of shillings, the cycle of life began a little more soberly and hopelessly once more.

Clerks, shop assistants and other white-collar workers led depressing shabby-genteel lives, and when sickness, unemployment or old age came on, waged a bitter struggle against starvation, but one seldom heard a word of complaint. At that time the copies of wills and other legal documents were made by scriveners in beautiful copperplate handwriting, for which they were paid at the rate of a halfpenny a folio (seventy-two words). A scrivener who worked hard for long hours could earn up to fifteen shillings a week at this skilled and exacting work. Many of them were able to marry and raise families (God knows how) on this meagre pay.

Clerks started life as office boys at five shillings a week for a fifty-four hour week. They ranked as middle-class people and aped the dress and manners of the rich, who in turn tolerated and treated them with a sort of patronising approval.

Members of the lower middle-class considered themselves superior to ordinary working people and would not mix with them socially. In their circles the word "class" represented something sacred. Middle-class men and youths maintained a ridiculous code of taboos. It was considered beneath their dignity to engage in any form of manuel work or even to carry a parcel in the street. If two friends met on a tramcar, an argument would ensue as who would have the honour of paying the fare, even if their combined capital only amounted to sixpence. Yet, with all their superficiality and acting, they were good and kind-hearted as a class, and if they had it, would hand out tuppence to a beggar as if they were plutocrats distributing largesse.

Bankers, industrialists, financiers, high-ranking members of the judiciary and administration lived in the select residential areas of Rathgar, Blackrock and Kingstown and in secluded mansions off the main road from Dublin to Bray.

The upper classes, as they liked to be called, were with few exceptions dedicated to the service of King and Empire. They were a closely knit community whose main interests centred around parliamentary reports from Westminster, the Money market, Foreign investments, gossip from the European Royal households, India, the movements of British troops and the current social activities of the Lord Lieutenant of Ireland.

The ruling class had little interest in the Irish people, its main objective being to colonise and exploit Ireland so that it would be a continuous source of revenue to the British Empire. The hierarchy of the established social order and State-sponsored religion maintained friendly relations with religious sects of all denominations, especially those that preached the Gospel "Peace on earth to men of good will."

Like the Pharisees of old, they proclaimed that all their aims and industrial activities were directed towards amassing wealth for the common good, but in reality they were perpetuating a wicked idealogy which decreed that two-thirds of the people should starve so that a privileged few could live in luxury.

The representatives of this select coterie patronised fêtes, garden parties and other entertainments held to raise funds for charity, making sure that if they ever did anything to help anyone, the whole world knew of their humane attitude and generosity towards others.

When in contact with the common people during the work-a-day round, the wealthy few hid behind a mask of aloofness, giving the general impression that they were so preoccupied with their own business worries that they didn't even know about the inhuman conditions under which the majority of their fellow-citizens existed.

The menace of poverty hung like a guillotine blade over the head of every breadwinner whose income depended on being employed by others. Kerr Quaney was no exception. He couldn't hope to save money from his salary. There seemed no way out of the economic morass in which he found himself, except to wait patiently for an opportunity that he could turn to good account.

Then one day, sitting in a tramcar, he overheard a conversation that set him thinking. A woman was giving out about how hard it was to catch a tram during the dark winter evenings:

"The black poles an' the white poles — y'can't tell one from t'other in the dark when it's rainin' cats an' dogs," she complained. "By the time y'find a white pole the tram's gone an' you're left there."

Kerr had the solution in a flash. Electric lights on the white stopping poles would soon solve that problem. But likely enough he wouldn't get a penny for his thought. Yet there seemed to be something more in the idea and he concentrated on it.

He worked hard until he invented an apparatus that he considered would be worth patenting. After scraping together all the money he could lay hands on he went down to the Patents Office and filed an application for protection for British Patent No. 15483 on the ninth of July, 1906, and the title of it was "Rocking mercury switch for Lamps."

The invention consisted of an enamelled sign shaped like a huge electric bulb. On the sign was a notice: "ALL CARS STOP HERE." The sign was designed to hang from a bracket on every white pole so that it would swing in the wind. Above the sign on each side of the bracket was an electric lamp, the switch of which was operated by a ball of mercury running in a tube. Every time the sign moved the lights blinked and drew attention to the stopping place.

The combination of sign and rocking mercury switch was used for years to indicate tram stopping places, until the Dublin Corporation improved the street lighting and flashing lights were no longer required to illuminate the signs. It was Kerr's misfortune, however, that another inventor took out patent rights for a similar device in England at the same time, so he derived no profit from his invention.

Despite these set-backs, Kerr had a driving ambition to advance himself and achieve some degree of financial security, but as things were going the prospects didn't look rosy. In Dublin of 1906, the onus was entirely on the individual. A man and his family were either swimming in the general current of affairs or submerged, and once down seldom rose to the surface again.

The general standard of education was low — not only in Dublin, but everywhere. The chancer and the quack flourished. The Sairey Gamp type of midwife ushered in the rising generation. In the Port of Dublin there were masters of ships of various nationalities who had scarcely enough education to write up a log book or sign their name, but who could be relied upon to take a ship to any part of the world. Travel and experience counted more than book learning. Life was a free-for-all scramble for the young and vigorous, and the reward of independence was well worth striving for.

This was the spirit of the times and these fanciful thoughts were running through the inventor's mind one Sunday morning until he snapped out of them and called his son, the second eldest:

"Joe, get your cap on and we'll take a run to Ringsend." The boy raced in — all excitement. "We'll be back around dinner time," Kerr announced, as they closed the front door.

It was a lovely morning. As they walked towards the tramstop Kerr took off his top hat and brushed it with his sleeve to polish the nap. Joe trotted along to keep up with his dad's brisk step.

"Daddy, how is it you're the only man in the street that wears a hat and coat like you do?" the child asked, looking up at his father earnestly.

"Well I suppose it does seem a bit queer to you," his dad replied as it dawned on him that he might look old-fashioned. "The moths have got on these anyhow. I'll have to buy an ordinary hat and coat" he added a bit regretfully.

When they arrived at Ringsend power station the boy's first impression was the vastness of the place, the smoke, smell and noise of machinery. Rows of bright electric lights were showing through the high windows although it was a fine sunny day.

"Look, Daddy, they forgot to put out the lights," said the boy, followed by "Gosh! Y'can climb right up to the top of those big chimneys and walk around on the platforms!!!" His gaze followed the long steel ladders running up to the dizzy height of the chimney cappings from which the grey smoke curled against a clear blue sky.

A puff of wind brought a strong smell of gas from the Gas Company's retorts on the far side of Ringsend Dock. Although it was Sunday morning the great coal grabs were unloading a ship, the rusty hinged buckets groaning and yawning as they opened and closed in their Herculean task of providing fuel for the furnaces.

At the top of the coal-tower, the wheels of tipper trucks screeched as they flashed along the runways to drop their loads into the fuel chutes. The station itself hummed like an enormous hive and as they entered the noise of many engines moderated to a steady drone.

Clusters of bluish lights, suspended from the roof, cast reflections on the vast red-tiled floor. Everything was spotlessly clean and polished.

Of the six neatly spaced machines, four were working to produce current for the normal Sunday morning load and the other two were idle — standing by— in case of emergency. The massive generators, linked up to quadruple the output of electrical energy, gave a fantastic display of synchronised mechanical action.

Pistons, activated by a tremendous pressure of superheated steam, thrust up and down purposefully. Conrods flashed, cam shafts oscillated at a perplexing speed and the great flywheels hummed contentedly.

High above all, near the main steam pipes, twin-balled governers whirled in concentric rings, like pawn shop signs caught in a cyclone.

Joe followed his father up the steel ladder to the platform that stretched full length of the main switchboard. A side door opening into a space behind the board had a gleaming skull and cross-bones painted on it.

"Daddy, what's that for?" asked the boy.

"It's a warning, Joe, look in here," his father replied, opening the door cautiously. In the space between the back of the switchboard and a long wall, a narrow boarded catwalk ran to another door at the far end.

"See the bare ends of the cables running up to the backs of the switches? They're alive. You've only to touch one and you're gone. That sign on the door means Death."

"And do you have to go in there, Daddy?" asked the boy, his voice a mixture of admiration and awe.

"Well, I suppose I would — if anything happened, — that's one of the risks of being an engineer," his father replied.

Joe said nothing more. His newly-acquired knowledge quietened him and made him feel prouder of his Dad. They walked along the platform fronting the main switchboard, the big ebonite panels crowded with clock-like instruments that had funny dials and ranks of big switches with shiny black handles that could "liven" or "deaden" a line at a pull.

"Don't forget it's Death if you touch anything around here," his father again warned as they passed another door with the skull and crossbones sign on their way to the engineers' offices.

Mr. Toule was in his room when they knocked and went in.

"Well, Kerr, how do you do? It's ages since we met," said his chief warmly.

"This is Joe, the second eldest, a chip of the old block," replied Kerr, introducing the boy. Then they got into quiet and earnest conversation.

"Suppose you heard I'm leaving the Company," said Mr. Toule.

"No!" replied his colleague, astounded.

"I am," confirmed his chief. "Feel my work is finished here and I'm off to America again."

Kerr listened in silence. All his built up hopes for his future with the Company vanished at the news.

"Glad to see you today, Kerr, because I'd like you to know that I've done all I could to further your interests, but the way it is now everything is done by the Board of Directors. New men are being taken on, promotions are made, executive positions are filled — you know the old story, friends of the directors and shareholders get all the plum jobs and no one can do a thing about it."

Kerr could see what had happened. The Company, started by a few enthusiastic and brilliant men who knew one another personally, had developed into a vast money-making machine, and in the manner of all machines was a soul-less entity that got out of touch with the pioneers who founded and developed it as the years passed by.

After successive Annual General Meetings it was apparent that the handful of tycoons who now controlled the Company, planned to pay a steady six per cent per annum to Preference shareholders (let Ordinary shareholders be thankful for what they got), and put the rest of the profits into vast reserve funds for future operations.

It never seemed to occur to these plutocrats that the Company owed something more to the conscientious workers of all ranks, who turned its initial capital outlay into a mine of wealth, than skimped pay-packets, promises, and occasional meritorious mention at Board Meetings and banquets.

"I brought up the subject of raising the engineers' salaries several times," continued Mr. Toule, "but the matter was always 'deferred for the time being.' The Board's view seems to be that permanent positions are hard to get, and those who are lucky enough to be in one, should be grateful for what they have. The Board also says that all-round costs are rising, and that if the exorbitant demands made by trades unions ever have to be met, the Dublin United Tramways Company may well find itself in the bankruptcy Courts."

Kerr sighed a little wearily, thinking well before he replied to his friend:

"I'd take most of what you said with a grain of salt, especially about the D. U. T. Co., winding up on the scrap heap, but as regards the refusal to raise salaries when we're all worried about the price of boot-leather and all that, all I can say is, a Limited Liability Company has no conscience."

Then — as if alerted by some telepathic tremor their glances met, suddenly and apprehensively — Joe was missing.

"Joe!" his father called, raising his voice above the hum of machinery, but there was no sign of the boy. They looked in the other offices and the toilet, then went along towards the switchboard platform. When they came to the first death's head door, Mr. Toule opened it slowly and peeped in.

"Oh God, Kerr look!" Joe was walking along the narrow gangway towards the far door, his back towards them.

"Sssh, don't call him," said Kerr — racing along the platform to the other door and opening it.

"Come on, Joe, don't touch anything," he pleaded as the boy stepped carefully towards him. A few more paces and he was in his father's arms.

"What on earth did you do that for, after all I told you?" asked his father when he recovered his composure.

"I wanted to be as brave as you, Daddy," replied the boy a little tearfully. The men were too relieved to question the lad further so they said, "Goodbye and God Bless," thankful for his escape.

The family had moved to a modern red brick residence in Glasnevin more in keeping with the style expected of a professional man. The house was conveniently situated on a main three-minute service line to Nelson's Pillar, the focal point of all Dublin's tram traffic and there was a tram stop on the footpath opposite the hall door.

A telephone from which any of the power stations or tram depôts could be called was installed on the Company's private line. In those days a telephone added prestige to anyone's house and Kerr's was the envy of all his neighbours. It was a great help to him in his work, and saved many a long and tedious journey when dealing with break-downs on far flung lines.

For the first few months the latest means of communication was a novelty and a blessing but, when his phone number became known to hundreds of tramway employees and their friends it became little short of a nuisance. Telephones were few and far between and when it became known where one was installed everyone wanted to use it. The old-fashioned bell that sounded like a fire alarm began to ring at all hours, sometimes about the most trivial matters.

People looking for jobs, urgent requests for a local doctor to be sent to a certain address, for an ambulance, the fire brigade or police, until the house was receiving more calls than the exchange in the G. P. O. Kerr Quaney was probably the first man in Ireland ever to say, "I wish I'd never seen or heard of telephones." Many a time he was tempted to disconnect the instrument and throw it out through the window, but a dedicated sense of duty caused him to leave it be and put up with the consequences.

One Saturday afternoon a call came from Ringsend power station.

"Come out quick, Mr. Quaney, there's something wrong with one of the engines. They can't stop it — it's terrible!" The speaker was almost incoherent with fear.

In half an hour Kerr was at the power station, and on entering, found the place deserted. He yelled at the top of his voice — no one answered. The floor of the station trembled as if an earthquake was imminent and smoke was rising around number three engine. The normal hum of the machine had changed to the menacing roar of a cyclone.

The thirty-ton flywheel raced like a buzz-saw. The smell of burning oil told him that the bearings must be hot — nearly red hot. The flywheel could burst any minute and wreck the station! No wonder everyone had fled from their posts.

Then through a break in the smoke he saw the cause of the trouble. The swing-ball centrifugal govener, that should regulate the pressure of steam to the engine by opening and closing the throttle had jammed, with the throttle full open.

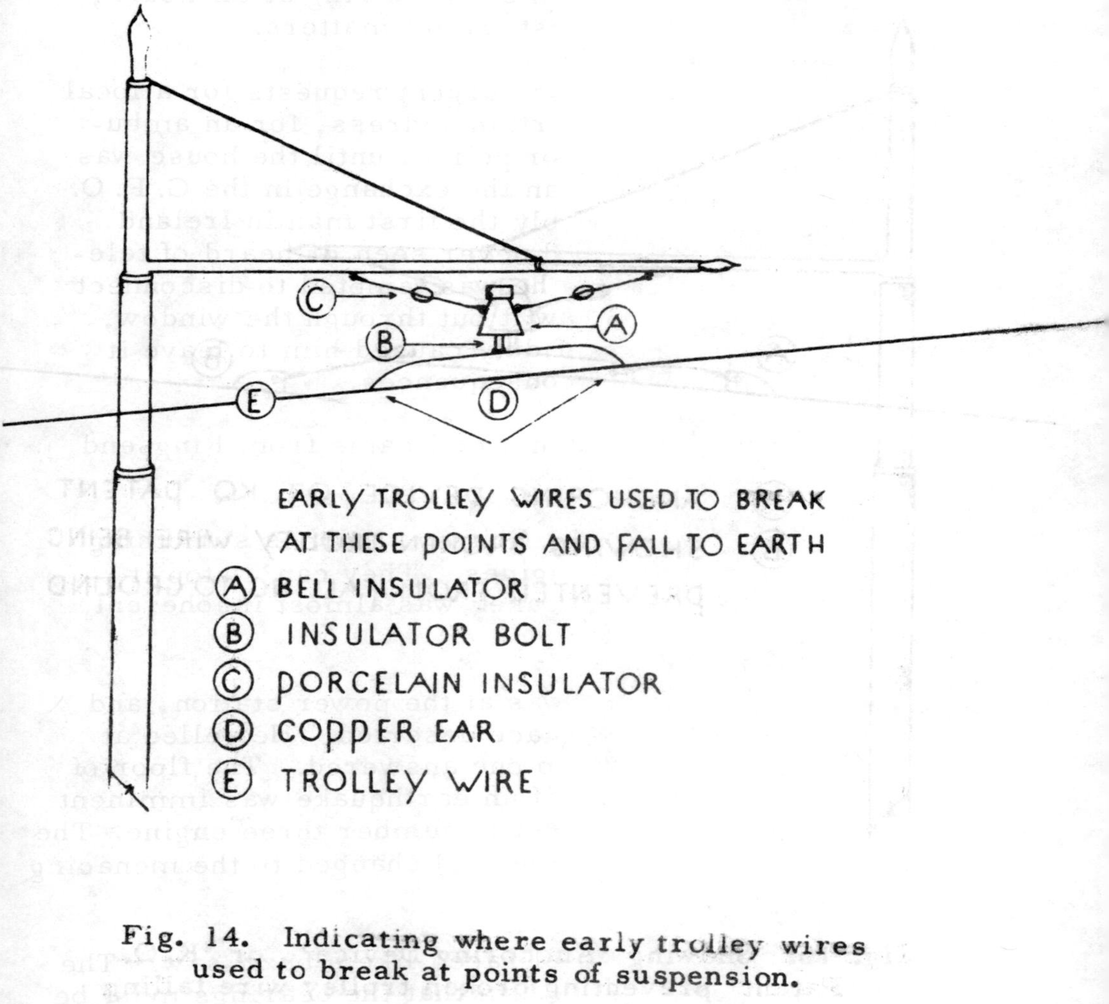

Fig. 14. Indicating where early trolley wires used to break at points of suspension.

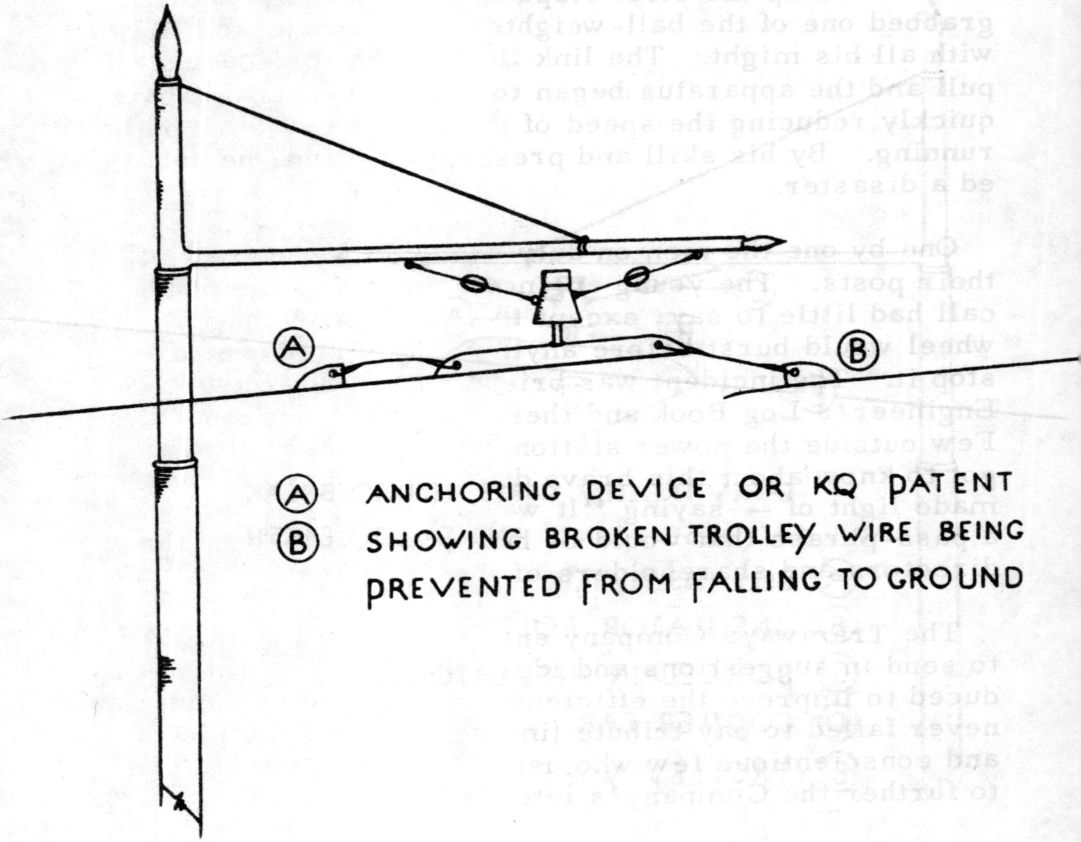

Fig. 15. Showing 'Anchoring Device', or 'K.Q. Patent' preventing broken trolley wire falling to earth and causing damage.

Kerr ran up the steel steps to the oiler's platform, grabbed one of the ball-weighted links and jerked it with all his might. The link flexed with the sudden pull and the apparatus began to spin and function again, quickly reducing the speed of the engine to its normal running. By his skill and presence of mind, he averted a disaster.

One by one the men on duty sheepishly returned to their posts. The young engineer who made the phone call had little to say, except that he thought the flywheel would burst before anything could be done to stop it. The incident was briefly recorded in the Engineer's Log Book and there was no more about it. Few outside the power station staff in Ringsend ever got to know about this brave deed which Kerr always made light of — saying "It was all in the day's work," a pass-phrase that would be heartily endorsed by the directors and shareholders of the Company.

The Tramways Company encouraged its personnel to send in suggestions and ideas that could be introduced to improve the efficiency of the service, and never failed to pay tribute (in words) to the faithful and conscientious few who used their creative ability to further the Company's interests.

In the early days of electrification, boys were employed at points where tram routes crossed, to switch the tracks so that a tram could move in its right direction and switch them back again for trams going in a different way.

This operation sometimes caused traffic delays and bad feeling among drivers and conductors, especially if for some reason or another the boy wasn't there with his lever to switch the points.

When the news circulated that one of the staff engineers had invented an electrically operated track shifter, Kerr went down to the Works at Inchicore to see the working model, which was nearly ready for a test.

The device was invented by an engineer named Tierney and the prototype (quite a hefty piece of work) was made by Malone, General Foreman at Ringsend. The invention finally emerged as the "Tierney/Malone Patent Track Shifter." The first model was installed at a "T" junction known as Dunphy's Corner where the Glasnevin and Phoenix Park tram lines intersected.

The shifting mechanism was controlled by a trip switch mounted on the overhead wires. The trolleys of passing cars moved a lever that closed the switch points and moved the tracks, and the apparatus automatically reset itself when the tram passed on its way. The electric track shifter was a great success from the time it was installed, and was eventually used at several points in the service.

At the time Kerr visited Inchicore, the men were still talking about an incident that made local history. The Tramway Works; a jumble of old machine shops and lean-to sheds from the old horse-tram days, was divided into two sections by a public road through which the trams passed on the way to their parking stations in the sheds.

At one of the main gate piers a long and a short wall joined at a point like the letter "V." In the angle of the "V" stood an outsize steel pole which carried a network of trolley wires leading to the sheds. A few feet from the pole was a double-gate through which men passed all day long in the course of their work.

One damp and dismal morning a mongrel strayed into the yard, went over to the pole and lifted his hind leg as any dog would. The instant he made contact there was a yelp that could be heard the length and breadth of the Phoenix Park, the poor brute sprang into the air and fell dead at the feet of two men who were carrying planks to the sawmill.

"God Almighty! That pole must be 'alive'," said one of the men.

"If it is, it must have been 'alive' for years," replied his mate, aghast.

The pole was tested and found to be "alive." After it was insulated and made safe, the men never tired of guessing the number of their fellow-workers who escaped the fate of the unfortunate dog.

Years passed by. The family increased in number but the family income remained static. The cost of living rose steadily, the economic situation became worse. There was only one consolation, Kerr was exempt from having to pay Income Tax; but as he often said dryly, he had no income anyway, his salary was all spent before he got it, so the family struggled on through a long period of hard times.

Then like a shaft of sunlight through the gloom, Kerr was inspired to action by another idea — one that if developed and patented, would surely turn out to be a money-spinner. Bound by the requirements of the Patent laws to keep his invention secret, he told nobody about it but set to work, searching and borrowing until he acquired a few electric cells, a vacuum pump, magnets, sheets of vulcanite and other sundries — enough as he thought to make a preliminary experiment. When the time came around for his annual holidays he was ready to start.

One glorious day in August, he was out in the back garden of his home arranging bits and pieces for an experiment. For the want of something better, he used the top of a large packing case as a work bench. Just as he was settling down to work, his son, a lad of nine, came out of the house and stayed around to see what was going on.

"Daddy, that man Blériot who flew over the sea in his aeroplane yesterday, isn't he an inventor?"

"He is, Joe, but run away and play, I'm busy," replied his father a little sharply.

"Why do you have to work so hard, Daddy? You always used to play with me, now you never do," the boy persisted.

"Oh, I've to work for you son, and Mammy and your little brothers and sisters; everyone has to work to keep the home going," his father replied softly.

"Well Daddy, if you invent something, will you become rich and famous like Mr. Blériot?" the boy continued unabashed. His father pondered awhile, looking at the pitiful collection of scrap and junk and the makeshift bench, on which he was trying to carry out a difficult experiment. Completely frustrated through lack of funds and proper equipment, he suddenly felt that he had come to the end of things and would have to give up trying. Then his eyes flared up in rebellion against the injustice of it all.

"Son," he replied, "some day when you grow up you'll hear old people say 'Everyone has a Cross to carry.' I think they're right," he concluded as he gathered up and put away for ever the components of his still-born invention on which he had built up hopes to win independence and security.

20

IN 1900, a rising spirit of resentment against being treated as slaves began to spread among the people of Ireland. Workers got together, compared notes and decided it was time to organise and fight for better conditions. Before the year was out, fifty thousand men and women had enrolled in trades unions or guilds.

The strike weapon appears to have been first used in Ireland in the great industrial city of Belfast and after some minor successes there, was introduced to Dublin. Organised workers in the building trade and Port and Docks Board, carried out a series of strikes but nothing much was achieved in the way of obtaining pay increases or improved working conditions, as the strikers' jobs were quickly filled by volunteers from the masses of the unemployed.

Between 1901 and 1908, the cost of living in Dublin went up by ten per cent. and the working class, goaded to desperation by the worsening economic conditions, began to pay more attention to trades union leaders who counselled them to make a stand and fight for better times, rather than die of starvation.

In the first decade of the century, trades unions had limited aims and ideals in comparison to what they have now-a-days. Each union strove to obtain more pay, shorter working hours and better conditions for its own members, and the dockers, carters, railwaymen, bricklayers and unskilled workers fought their own battles in their own particular sphere of industry. Generally, the early strikes were one-sided contests. In most cases employers knew the odds were in their favour because the workers would soon be forced to surrender and return to work, from sheer pressure of starvation.

From a business-man's point of view, Dublin offered tempting opportunities for capital investment. The whole city was ripe for exploitation — a shrewd industrialist could scarcely lose. An army of skilled and unskilled labour was clamouring for employment. Thousands of young men were willing to work for five shillings a week, factory girls worked at all kinds of jobs for three and six a week, domestic servants worked fourteen hours a day for half a crown a week and their "keep" in middle class homes. Cheap labour — the untapped source of unlimited wealth — was there in abundance, waiting to be harnessed to industry.

The slave rates of pay were generally reflected in the living conditions of the workers. Between 1905 and 1908 an average of fifty thousand unskilled workers and their families, lived in rooms that were unfit for human habitation. More than a hundred thousand depended on these for their daily bread.

Jails, hospitals and lunatic asylums were overcrowded. The death rate 27.6 per cent in the Dublin slums was the highest in all Europe. Most of the landlords who owned slum property lived permanently in England, and had their rents collected by Irish agents.

But if the unskilled masses were reduced by an unjust and heartless society to living in habitations little better than animals' dens, the lower-middle-class workers had their own kind of purgatory to endure. This semi-skilled class was the bulwark of religion and but for it the entire social edifice would have crumbled and collapsed.

The accepted philosophy of the middle class was that poverty was the foundation of Christian perfection. It was man's heritage to suffer and it was up to them to show others how to bear it. By denying themselves all luxuries and amusements, the majority of middle-class people managed to live in rented houses a cut above the tenements — those who gave way to drink soon found refuge in a pauper's grave.

Thousands of lower-middle-class families dragged out an existence with insufficient fuel or clothing and suffered from lack of bodily needs for a whole lifetime, before succumbing to endemic diseases. Mothers of growing families, worn out by domestic cares and worries long before their time, sank into torpor and ended their lives in what was conveniently described by medical men of the period as a "decline."

Social workers who studied the existing economic problems, maintained that the grossly unfair distribution of wealth was not entirely caused by Irish employers exploiting Irish workers, but partly because the people of Dublin had through indirect taxation to support an army of highly paid State officials, military garrisons, a host of corporation officials, as well as the Metropolitan Police force and the Royal Irish Constabulary.

Then, without any fanfare of trumpets, flag-wagging or red carpets, James Larkin, a florid, formidable-looking Liverpool-Irishman, stepped off the boat at the North Wall early in 1906, when the mass misery of the Dublin workers had reached a hitherto unknown depth.

Jim had known somewhat similar conditions in his own home town, and had dedicated his life to fighting for the working class in any part of the world. His immediate aim was to organise the dockers and unskilled labourers in the Irish ports, and he had the full sympathy and backing of the cross-channel unions in his venture.

Within a few months the National Union of Dock Labourers was formed, and Larkin's imposing figure dominated the Liffey waterfront. His dynamic personality, sincerity and utter selflessness, attracted men, roused them from apathy and inspired them with hope for the future.

Brimming with confidence, he called the dockers out on "lightening" strikes that made shipowners feel as if they were shipwrecked. Every strike was well timed and successful. The members of his union gained one concession after another, more pay, better hours and bonuses for discharging ships in time to sail on the following tide. Jim Larkin then went off to America to study trades unionism in that country.

The successes of the Docker's Union caused a wave of unrest among the working class in Dublin. Forced to work for wages that would scarcely buy the necessities of life, more workers joined the established unions and braced themselves for the struggle for a living wage.

As economic conditions steadily worsened, relations between workers and employers deteriorated and sporadic strikes took place in firms all over the city. As a counter-move announcements appeared in the newspapers that many of the firms involved in labour disputes were making arrangements to bring workers over from Britain to take the places of those on strike.

Some of the strikes called by unions that catered for skilled tradesmen were successful, and new and improved rates of pay were gained by men who could not easily be replaced by strike breakers.

The Dublin Federation of Employers, realising that the challenge to their interests and authority was becoming stronger every day, held frequent meetings to find ways and means of frustrating the demands of the unions. The whole industrial situation was becoming chaotic and beginning to get out of hand, when Jim Larkin returned from the United States in 1908.

During his stay in America, Jim had apparently been indoctrinated with the principles and teachings of the Industrial Workers of the World. His outlook had changed. No longer was he content with the simple socialist plan for individual trades unions to fight for the betterment of their members; he now favoured the Syndicalist view that socialism was out of date and undesirable.

This extremist group sought to bring about a social revolution by securing control of industry in all countries, and the complete removal of the capitalist class. His mission now was to expose the abuses of the capitalist system that were being perpetuated by the ruling class in Dublin, stoke up the fires of resentment, and condition the minds of the proletariat for an all-out class war.

A great number of Larkin's rag-taggle cronies and comrades in the Docker's Union were too ignorant to understand or appreciate his latest advanced theories about taking over control of industry and forming a Workers' Government — but they thought him a dedicated genius, and acclaimed him as director of the working class.

Larkin's advanced ideas, however, were not wholly accepted by the councils of small but long-established trades guilds, whose basic principles were to strive for the betterment of their members in their own little sphere of industry. They could foresee that if Larkin rose on a wave of popularity and was swept into power all unions would come under his dictatorship. Individual unions would be denied the democratic right to shape their own policies and fight their own battles and worse still, the administrators of the guilds would probably lose their jobs.

Despite the suspicions and jealousies of his contempories, Jim Larkin went ahead with his plans and formed the Irish Transport and General Workers' Union, with headquarters at Liberty Hall. From the time of its inception, Larkin became front page news in the Dublin papers, especially the "Irish Independent" and "Evening Herald," owned by William Martin Murphy, who was also head of the Dublin Federation of Employers and Larkin's bitterest enemy in the ensuing class war.

The newspapers repeatedly pointed out that the Syndicalist labour movement which Larkin represented and spoke for, was diametrically opposed to the teachings and doctrines of all Christian churches. Although harassed by a steady stream of adverse and sometimes acrimonious propaganda, Larkin continued to hold meetings and stimulate the crowds with his fiery oratory.

The riff-raff of the city listened and made him their idol. Embittered men who had lost employment in union disputes, defeated strikers, the homeless who hadn't even a tenement roof over their heads, all those who were engaged in a hopeless struggle for survival rallied around his platform and managed to scrape together or beg a few pence a week, to hand over as a contribution to his union.

In exclusive clubs and dingy pubs, Larkin's name was on everyone's lips. Even in the schools the boys and girls were singing:

"Woncha join oh wontcha join,
Oh wontcha join Jem Larkin's union?"

One dark and bitterly cold evening in November, 1912, a crowd began to gather at the junction of Upper and Lower Sherrard Streets, where five roads meet. Groups came from the stinking tenements of Dorset Street, Summerhill and Middle Gardiner Street. It would be hard to imagine a more tattered, woebegone batallion of scarecrows, gathering to make a protest against injustice and starvation. A big rally and torchlight procession to Liberty Hall was about to take place.

Union officials moved around trying to sort out the band and keep it in its place. There was an air of secrecy; everything was being done quietly. Joe Quaney, infected by the enthusiasm of his school mates, had obtained permission from his father to watch the procession, but on no account to take part in it.

The band was mustered at last and suddenly began to march off without anyone giving an order. Hundreds of men and boys carrying broom handles and clothes-line poles to which tin cans full of rags were attached, fell in six deep and marched behind the band. The ranks were irregular, everyone walked at his own pace.

Nearly half the marchers were messengers and boys who sold William Martin Murphy's newspapers on the streets. These ragged, starry-eyed urchins, trying to look more than their age, dodged into the ranks, expertly avoiding the boots of the grown-ups as they smacked their bare feet hard on the freezing stone setts every now and then to keep the circulation going. Although starvation was written on their pinched, pallid faces, not a whimper of complaint was heard from any one of them.

When the procession reached Jones's Road, the band automatically slowed down to a halt, broke formation and rushed in a body over to Fannin's pub on the corner, where some of the members evidently had slate accounts. The whole procession would have followed, but the place was crowded in a few minutes and the frustrated ones had to wait outside, envying those who had gained admittance to the warmth of the bar.

During the interlude, men with bottles went around pouring paraffin oil on the rags in the torch cans and when all were filled and waiting, there were angry shouts of "Gettam outa th' pub; gettam outa th' pub or we'll all go home again."

At last, the message got through to the band and the musicians returned to their places well primed with porter. A preliminary bump on the drum and discordant piping of flutes indicated that they were ready for action.

"Bloody shame wastin' money on paraffin oil when we could be gettin' good porter for it," grumbled someone as the matches were passed around for lights. Then, inspired by the flare of a hundred torches, the band struck up.

Drums thundered, cymbals crashed, tin whistles, flutes and mouth-organs tried to outdo one another as the procession moved off to a pathetic rendering of the anthem:-

"God save Ireland, said the heroes,
God save Ireland, say we all,
Whether on the scaffold high,
Or on the battlefield we die,
Oh what matter when for Erin dear we fall."

The half-tipsy band played at full blast and the smoky torches flared, making a brave show against the murky panorama of gaunt stone buildings as the ragged revolutionaries marched down Buckingham Street, turned into Amiens Street and swept along under the loop-line railway bridge into Beresford Place, where Jim Larkin was addressing a meeting.

A great cheer went up when the new contingent arrived and swelled the crowd to imposing proportions. Big Jim Larkin took advantage of the interruption to get a better footing on the seat of the outside car he used as a stand when addressing a crowd at Liberty Hall. Around him was a phalanx of hand-picked stalwarts, every bit as desperate and deadly as the bloodthirsty mob that stormed the Bastille and created a new social order in the name of Liberty, Equality and Fraternity.

"And I'm goin' to talk to you tonight about some of the industrial Shylocks in this city that's squeezin' the last ounce outa the unfortunate men, women an' children they've enslaved."

The speaker continued when the noise subsided:

"You — the working people of Dublin — are bossed by a gang known as the Employer's Federation, but what I call the Employer's Secret Society. And the head of it is William Martin Murphy of the Dublin United Tramways Company.

"This fellow fancies himself as a dictator of industry, owns newspapers, hotels and the biggest drapery business in Ireland. Controls tramways and railways all over the place. This Saviour of the People (according to his own newspapers), is a multi-millionaire.

"And how does he make his millions?" he asked in a carefully modulated voice, so as to let the words sink in. "He makes 'em outa you an' th' likes o' you and all because you don't stand together and fight him," he roared, making one of those theatrical gestures that marked him out as the greatest mob-orator of all time.

The crowd cheered and shouted encouragement.

"Good ole Jim — Stuff to give 'em, Jim . . . Give 'em more!"

"These capitalists are keeping wages down by employing scabs and blacklegs in place of those on strike; so as to starve the strikers into submission. There's forty thousand hungry unemployed in this city today, and I'm sure there isn't one of you here that hasn't seen his own kith and kin die, as if they were living in the time of the Famine.

"But there's no famine in Dublin today except the one that's artificially created by your so-called Christian employers, and the mealy-mouthed pack o'hypocrites that backs 'em up. You're starving in the midst of plenty, and we're going to show this rotten city that, when Christ said 'The Kingdom of God is within us' He meant all of us, and not just the fellows in the frock coats."

When the applause died down the speaker continued:

"You're all praying from morning till night that something'll happen and you'll get rises in wages or better jobs, but you oughta know by now what to expect from employers. They'll just keep the pressure on until they squeeze life itself outa you. There's only one way to get anything out of employers and that is" — his voice rose to a crescendo of fury as he roared like an old-time general launching an attack — "Get up! Get up! GET UP OFF YOUR KNEES AND FIGHT!"

A prolonged burst of cheering followed and hats went up in the air. The old horse dozing in the shafts of the hackney car got such a fright at the tumult that it sprang to attention, jerked the traces and threw the speaker off his balance, almost toppling him on to the heads of his friends and admirers.

Young Joe, shocked at the speaker's blasphemous reference to prayer, burrowed his way through the crowd like a ferret, and ran all the way home to tell them about the procession and the crowds that were flocking to join Jim Larkin's Union.

A great number of workers were impressed and influenced by that fighting speech, and the membership of the Irish Transport and General Workers' Union rose steadily, its policy became more aggressive and the cult of "Larkinism" spread. Large-scale employers of labour such as the Dublin Port and Docks Board, timber and coal importers and the cross-channel shipping companies were all beset by strikes.

James Connolly, Secretary of the I. T. & G. W. U. in Belfast, assured of the backing of his comrades, began to take a more active interest in Union affairs in Dublin. Connolly was a moderate type of socialist who wanted Labour to get into power in a democratic way, by gaining votes and representation in Parliament and, although he co-operated with Larkin in expanding the I. T. & G. W. U. he disliked him and said more than once in public that he considered Larkin unreliable and dangerous.

The syndicalist ideology of Larkin and the form of socialism advocated by Connolly were in conflict with the basic principles of Christianity, and were denounced by Catholic authorities on social philosophy. As a result, a large section of Dublin workers would have nothing to do with "Larkinism!"

Surrounded by the rising turbulence of class consciousness, conflicting ideals and industrial unrest, the pious people of all denominations maintained an attitude of benign tolerance and neutrality. In accordance with normal religious practice they were concerned more with the maladies of the soul than temporal affairs, but when some of Jim Larkin's utterances began to circulate in the inner sanctums and select vestries they were forced to take sides. The Churches lined up in a united front against Larkin and Connolly, with their threat to the sanctity of private property and the social order.

The clergy were horrified to think that any Dublin workers would listen to a man who wanted them to rebel against the teachings of the Church and denounced Larkin as a Marxist, an atheist, and some went so far as to refer to him as Anti-Christ.

So the wave of denunciation spread and in contrast his opponent William Martin Murphy, business potentate who employed an army of men, was raised to the status of a saint.

"God help poor Mr. Murphy, he deserves everyone's sympathy," was heard on all sides. "To think of all the employment he's giving and instead of being thankful they're all against him — the wretches."

The Church propaganda machine went into action and the Irish Transport and General Workers'Union was outlawed. The feeling of animosity between Church and Labour intensified and caused clashes in many parts of the country. When James Connolly, the 1916 patriot/martyr to be, went to address a meeting of friends and supporters in Kilkenny, the Men's Confraternity turned out in force and chased him out of town.

In 1912 James Larkin was elected to the Dublin Corporation, and the Irish Transport and General Workers' Union gained prestige and a certain amount of stability. Larkin, now the undisputed leader of the unskilled masses in Dublin, polled a vast number of votes and looked like becoming a political big-shot in the near future.

His rise to civic responsibility put his followers in good fettle, gave them confidence and ushered in a new series of strikes, none of which proved conclusive, but collectively prepared the ground and conditioned the minds of the people for the major clash between the forces of capital and labour, that everyone in Dublin knew was coming.

At this stage the red badges of the Irish Transport and General Workers' Union appeared on the lapels of the conductors and motormen in the Dublin United Tramways Company, and Kerr Quaney was approached by key men in the Labour movement to know if he would act as mediator between the men and the Company. The menhad great faith in him as he understood the hardships of their employment and always had a sympathetic ear for their grievances.

This offered to open up a new and potentially lucrative career, but Kerr reluctantly turned it down saying:

"I'm with you in many things, boys, but can't you see that a man in my position with my responsibilities must be a Company's man!"

All through the spring of 1913 the labour unrest in Dublin became more widespread, and more workers were influenced by Jim Larkin's inflammatory speeches. The Federation of Employers, headed by William Martin Murphy, autocrat and industrial dictator, agreed that Larkin and Connolly's Irish Transport and General Workers' Union must be crushed, as the first step towards securing industrial peace.

At a meeting in the offices of his newspaper the "Evening Herald," Murphy announced that any one of his employers who joined the Irish Transport and General Workers' Union would be instantly dismissed without an explanation. A few days later most of the employees in the Tramways Parcels Express were sacked. On the following day, August 22nd, the tram men were warned that if they went on strike for the reinstatement of their fellow-workers, the strike would not last a day as the Company had a skeleton staff to take over their duties, and had guarantees of protection from the forces of the Crown.

This ultimatum against the rights of workers to join any trade union they wished had violent repercussions, and a remarkable effect on public opinion. Up to this point, William Martin Murphy was regarded as one of the best of a bad lot of employers, but this edict confirmed him as a despot who wanted to over-rule the democratic rights of individuals. He also refused to recognise or negotiate with the workers' chosen representatives. The entire trades union movement lined up against William Martin Murphy overnight, and precipitated a crisis.

On the morning of August 26th, 1913, the smouldering fires of animosity and class hatred flared up and spread revolution mania among the working people when, at 10 a.m. the motormen and conductors stopped their trams and abandoned them on the streets of Dublin without a word of warning to the passengers.

Before the strike was a day old, an organised army of strike-breakers took over the trams, and services of a kind were maintained. The strikers wanted trouble and went out of their way to look for it. Trams were overturned and riots broke out all over the city. Scabs — men who refused to take part in the strike for some reason or another — were dragged off the trams and beaten up. Paving stones and stone setts were dug up and hurled at the trams. People became afraid to travel in them.

The Dublin Metropolitan Police, taunted and ridiculed for years in the columns of the "Irish Worker" turned on their tormentors and behaved like hoodlums. The blood-lust was on them, as with drawn batons they smashed down any man, woman or child who by sign, slogan or innuendo let it be known they were "Larkinites."

When the long threatened strike started and began to disrupt the normal life of the city, employers of labour decided it was time to force the issue. Four hundred members of the Employers' Federation pledged themselves to lock out any employees who refused to relinquish membership of the Irish Transport and General Workers' Union. Forty trades unions immediately took up the challenge and gave the strikers their wholehearted support and sympathy. This was the start of the great lockout, and within a few weeks forty thousand men, women and their families faced starvation.

The labour unrest dragged on into the winter of 1913 and life in the city became chaotic. The strike and strikers were condemned on all sides, the principals in the struggle were looked upon as enemies of the people. William Martin Murphy, hitherto upheld by pious people as everything a good man ought to be, was thrown off his pedestal and like a plaster saint, smashed into a thousand smithereens.

"He's dirtied his bib this time," said his enemies gleefully as they visualised Murphy's industrial citadel tumbling down and burying him in its ruins.

Larkin with his big talk and irresponsible promises was equally condemned, for reducing the condition of the poor to a worse degree of famine and despair.

"He'll burn in Hell for it," said the pious people, who could only weep and pray for those whose sufferings were caused by the insatiable greed of their masters.

English trade unionists were so moved by the famine conditions imposed on their brethren by the class war, that they loaded an old ship named the 'Lady Jocelyn' with food and sent it over to help the strikers fight starvation. They also offered to give temporary board and lodging to the children of the strikers in their own homes, but the Confraternities in Dublin objected to this form of relief and prevented the children being sent to England.

In contrast to these humanitarian gestures, Irish newspapers appealed to their readers to harden their hearts against the starving children of the strikers begging in the streets, as any charity extended towards them would only help to prolong the strike.

The "Irish Worker," official news sheet of the Irish Transport and General Workers' Union, played a prominent part in the wordy battles that arose out of the great Lockout. Its reporters had no inhibitions. They just wrote down what they thought of people, untroubled by laws about libel, defamation of character or the stroke of a censor's pen. Everyone knew that Jim Larkin's paper hadn't enough money to pay damages if a lawsuit went against it, so the journal was free from claims of any kind.

Names of firms that were involved in labour disputes were printed in big-letter capitals and all "incidents" arising from strikes were covered. Its scouts found out the names of policemen who handled "Larkinites" roughly and published them in a weekly blacklist. No citizen, Government or Corporation official from the Lord Lieutenant down to the lowliest blackleg scavenger, was safe from being pilloried in its columns.

The "Irish Worker" had a particular grudge against young male clerks who worked for low wages and kept up snobbish appearances. Hundreds of these were employed at from five shillings to seven and six a week, and could be seen apeing their masters in the swank saloons and select bars any Saturday evening. How they managed to do it on the money was one of the profound mysteries of the Dublin economy.

These "Limpets" were abominated by trades unionists, who regarded them as would-be bosses who could never be relied upon to take the workers' side in a class war. Lists of their names and addresses and the salaries they worked for appeared frequently in the "Irish Worker" to embarrass and belittle the unfortunate pen-pushers, who lived behind a screen of bluff to hide their poverty and sense of inferiority.

During the eight months of the strike, the Dublin United Tramways Company maintained services but incurred great losses and expense. Policemen stood on the platforms of trams driven by "blacklegs."

Crowds of "Larkinites" boarded the trams during rush hours and refused to pay fares. Conductors were assaulted so often, that after the first month of the strike they didn't press for fares when passengers refused to pay.

During those wearying months of tension and strife, the engineering and maintenance staffs of the Company worked incessantly, and often at the risk of their lives, to keep the services going, as murderous attacks were made on employees going home from work.

James Connolly's introduction of large-scale "sympathetic" strikes brought industry almost to a standstill, and mob gatherings became more frequent and violent. Coal carts were overturned in the streets and their contents pilfered. Delivery waggons were looted. Bakers' vans were held up and the bread distributed. Provision shops were broken into and ransacked. The mobs were beginning to take over when William Martin Murphy called on his friends in Dublin Castle to re-establish law and order.

Although martial law was never proclaimed, squads of infantry armed with rifles began to patrol the streets, and detachments of cavalry dispersed crowds. When the British trades unionists heard that the strike in Dublin was threatening to develop into a revolution they withdrew all help and support, and the greatest working-class revolt in the history of Ireland began to simmer down. The strikers, beaten by hunger and adverse public opinion, returned to work on terms dictated by their employers.

It was a clear-cut victory for the employers and others who specialised in making fortunes out of the sweat and agony of the poor, but Larkin and his colleagues achieved something infinitely greater — a moral victory — the like of which has never been eclipsed.

By his stark and inspiring call to action, Larkin showed the Dublin workers the power of combined effort and led them from obscurity and slavery into the corridors of organised representation, negotiation and conciliation. He aroused the public conscience, formed a new concept of the relationship between master and man, and upgraded the workers to a level of social recognition from which they have never descended.

The end of the strike marked the beginning of a new era for the rank and file of the Dublin United Tramways Company, but the handful of engineers and executives who planned and constructed the great modern transport system and remained faithful to their commitments throughout the period of trial and tribulation, had no share in the new deal.

To them, it was a matter of starting in to repair the wreck of what had been a vast and lucrative enterprise. Gone were the prospects of increased remuneration and prosperity — years of further sacrifice and retrenchment stretched before them. Faced with such a bleak future many of the original staff left the Company to seek careers where the outlook was brighter.

Kerr Quaney, dedicated as ever to his work, remained at his post, but like most of the old hands never fully recovered from the effects of the strike.

Within a few months the clouds of the 1914 War with their portent of industrial and financial chaos were piling up on the horizon, filling the minds of those who had to earn their living with fear and misgivings, and effectively putting an end to all wishful thinking for the future ...

Years later, when the chaos caused by strikes and war became things of the past, and the Company once again began to pay fruitful dividends, a crowd of eight-per-cent. Preference shareholders (who had weathered both storms without the loss of a penny), gathered in a lounge bar after a more gratifying than usual Board Meeting. In the interchange of hazy reminiscences a loud voice enquired:

"Whatever happened to that engineer — Quaney, man who invented the K. Q. patent? Always thought a man with his brains would be up here with us some day."

"Oh, he's one of those who's always thinking about his work, never mixes around much," replied his colleague. "What are you going to have?"

But Kerr Quaney, the dynamic engineer of the change-over days whose technical skill and invention of the "Anchoring device" and other patents which helped to make the Company pay its way, had lost his early ambition and withdrawn into a habit of life which he could not change. Resigned to adversity, he never became embittered or envied those who achieved wealth and fame.

It was one of Life's complexities that others should profit from his inventions, and that his reward should be just a frugal living and the joy of creative achievement. The flame of his genius burned brightly for brief spells before he passed on to join those immortals who, by hard work, self-sacrifice, patience and perseverance produced devices which helped to make the world a better place for everyone.

When the Provisional protection which covered the manufacturing rights of the "Anchoring Device" or "K. Q. Patent" lapsed after seven years, the inventor, for some reason, was unable to proceed with an application for full Patent Rights, so the invention, no longer protected by law, could be used by anyone. The era of rapid transit had arrived, and engineers used the device freely on tramway systems which were being installed in progressive cities and towns.

When the legal protection period expired, the popular name for the device, the "K. Q. Patent" died with it and was superseded by a technical term known to electricians as "Looping."

This safety device can be seen to the present day protecting wires and cables on telephone and telegraph poles on cross-country pylons the world over.

* * *

Volturna Press

With the present high printing costs, many good books are never published because it would not be economic to do so. Our aim is to publish such books in small editions at a reasonable price. This means that it is difficult to use ordinary book printing, and many of our books will be produced (as this one is) by offset litho.

We aim to produce all sorts of books which have a minority appeal, but have an especial interest in the fields of religion, education and conservation.

You can help by sending in an application for our current list, which will cost you only the price of a postage stamp.

We are very willing to consider all manuscripts offered to us for publication. Although we are located in Ireland, our activities are by no means limited to that country, and we are willing to consider manuscripts from — and to sell books in — every country.

Personal memoirs

The very best form of memorial for someone who has recently died is a short book, descriptive of his/her life, which may contain personal letters, diary extracts, etc. Such a book may be produced at a very small cost for distribution among friends (if necessary, only in two or three copies). We can do this economically and attractively. Write for details.

Volturna Press
Portlaw
Co. Waterford
Eire.